T0091564

Tools to Help Your
Children Learn Math

Strategies, Curiosities, and Stories to
Make Math Fun
for Parents and Children

Problem Solving in Mathematics and Beyond

Print ISSN: 2591-7234
Online ISSN: 2591-7242

Series Editor: Dr. Alfred S. Posamentier
Distinguished Lecturer
New York City College of Technology - City University of New York

There are countless applications that would be considered problem solving in mathematics and beyond. One could even argue that most of mathematics in one way or another involves solving problems. However, this series is intended to be of interest to the general audience with the sole purpose of demonstrating the power and beauty of mathematics through clever problem-solving experiences.

Each of the books will be aimed at the general audience, which implies that the writing level will be such that it will not engulfed in technical language — rather the language will be simple everyday language so that the focus can remain on the content and not be distracted by unnecessarily sophiscated language. Again, the primary purpose of this series is to approach the topic of mathematics problem-solving in a most appealing and attractive way in order to win more of the general public to appreciate his most important subject rather than to fear it. At the same time we expect that professionals in the scientific community will also find these books attractive, as they will provide many entertaining surprises for the unsuspecting reader.

Published

Vol. 10   *A Problem Solving Approach to Supporting Mathematics Instruction in Elementary School: A Guide for Parents, Teachers, and Students*
by Sheldon Rothman

Vol. 9   *The Mathematics Coach Handbook*
by Alfred S. Posamentier and Stephen Krulik

Vol. 8   *Tools to Help Your Children Learn Math: Strategies, Curiosities, and Stories to Make Math Fun for Parents and Children*
by Alfred S. Posamentier, Gavrielle Levine, Aaron Lieberman and Danielle Sauro

Vol. 7   *A Central European Olympiad: The Mathematical Duel*
by Robert Geretschläger, Józef Kalinowski and Jaroslav Švrček

For the complete list of volumes in this series, please visit www.worldscientific.com/series/psmb

**Problem Solving in
Mathematics and Beyond** Volume **08**

Tools to Help Your Children Learn Math

Strategies, Curiosities, and Stories to

Make Math Fun

for Parents and Children

Alfred S. Posamentier

New York City College of Technology, City University of New York, USA

with

Gavrielle Levine
Aaron Lieberman
Danielle Sauro Virgadamo

Long Island University, USA

World Scientific

NEW JERSEY · LONDON · SINGAPORE · BEIJING · SHANGHAI · HONG KONG · TAIPEI · CHENNAI · TOKYO

Published by

World Scientific Publishing Co. Pte. Ltd.
5 Toh Tuck Link, Singapore 596224
USA office: 27 Warren Street, Suite 401-402, Hackensack, NJ 07601
UK office: 57 Shelton Street, Covent Garden, London WC2H 9HE

Library of Congress Cataloging-in-Publication Data
Names: Posamentier, Alfred S., author. | Levine, Gavrielle, author. |
    Lieberman, Aaron, author. | Sauro, Danielle, author.
Title: Tools to help your children learn math : strategies, curiosities, and
    stories to make math fun for parents and children /
    by Alfred S. Posamentier, Gavrielle Levine, Aaron Lieberman, Danielle Sauro.
Description: New Jersey : World Scientific, [2018] | Series: Problem solving in
    mathematics and beyond ; Volume 8
Identifiers: LCCN 2018024554| ISBN 9789813271425 (hardcover : alk. paper) |
    ISBN 9789813272477 (pbk : alk. paper)
Subjects: LCSH: Mathematics--Study and teaching--Parent participation.
Classification: LCC QA135.6 .P6482 2018 | DDC 372.7/044--dc23
LC record available at https://lccn.loc.gov/2018024554

British Library Cataloguing-in-Publication Data
A catalogue record for this book is available from the British Library.

For any available supplementary material, please visit
https://www.worldscientific.com/worldscibooks/10.1142/11019#t=suppl

Typeset by Stallion Press
Email: enquiries@stallionpress.com

Printed in Singapore

About the Authors

Alfred S. Posamentier is currently Distinguished Lecturer at New York City College of Technology of the City University of New York. Previously, he was the Executive Director for Internationalization and Sponsored Programs at Long Island University, New York. This was preceded by a five-year period where, he was Dean of the School of Education and tenured Professor of Mathematics Education at Mercy College, New York. He is now also Professor Emeritus of Mathematics Education at The City College of the City University of New York, and former Dean of the School of Education, where he was tenured for 40 years. He is the author and co-author of more than 60 mathematics books for teachers, secondary and elementary school students, and the general readership. Dr. Posamentier is also a frequent commentator in newspapers and journals on topics relating to mathematics and education.

After completing his B.A. degree in mathematics at Hunter College of the City University of New York in 1964, he took a position as a teacher of mathematics at Theodore Roosevelt High School (Bronx, New York), where he focused his attention on improving the students' problem-solving skills and at the same time enriching their instruction far beyond what the traditional textbooks offered. During his six-year tenure there, he also developed the school's first mathematics teams (both at the junior and senior level). He is still involved in working with mathematics teachers and supervisors, nationally and internationally, to help them maximize their effectiveness. During this time he earned an M.A. degree at the City College of the City University of New York in 1966.

Immediately upon joining the faculty of the City College in 1970, he began to develop inservice courses for secondary school mathematics teachers, including such special areas as recreational mathematics and problem solving in mathematics. As Dean of the City College School of Education for 10 years, his scope of interest in educational issues covered the full gamut educational issues. During his tenure as dean he took the School from the bottom of the New York State rankings to the top with a perfect NCATE accreditation assessment in 2009. He achieved the same success 2014 at Mercy College, which received both NCATE and CAEP accreditation during his leadership as Dean of the school of education.

In 1973, Dr. Posamentier received his Ph.D. from Fordham University (New York) in mathematics education and has since extended his reputation in mathematics education to Europe. He has been visiting professor at several universities in Austria, England, Germany, Argentina, Turkey, Czech Republic, and Poland, while at the University of Vienna in 1990 he was a Fulbright professor.

In 1989 he was awarded an *Honorary Fellow* at the South Bank University (London, England). In recognition of his outstanding teaching, the City College Alumni Association named him *Educator of the Year* in 1994, and in 2009. New York City had the *day*, May 1, 1994, named in his honor by the President of the New York City Council. In 1994, he was also awarded the *Grand Medal of Honor* from the Republic of Austria, and in 1999, upon approval of Parliament, the President of the Republic of Austria awarded him the title of *University Professor of Austria*. In 2003 he was awarded the title of *Ehrenbürger* (Honorary Fellow) of the Vienna University of Technology, and in 2004 was awarded the *Austrian Cross of Honor for Arts and Science, First Class* from the President of the Republic of Austria. In 2005 he was inducted into the *Hunter College Alumni Hall of Fame*, and in 2006 he was awarded the prestigious *Townsend Harris Medal* by the City College Alumni Association. He was inducted into the New York State *Mathematics Educator's Hall of Fame* in 2009, in 2010 he was awarded the coveted *Christian-Peter-Beuth Prize* in Berlin, and in 2017 he received the *Summa Cum Laude nemine discrepante* Award from Fundacion Sebastian, A.C. in Mexico City.

He has taken on numerous important leadership positions in mathematics education locally. He was a member of the New York State Education Commissioner's Blue Ribbon Panel on the Math-A Regents Exams, and the Commissioner's Mathematics Standards Committee, which redefined the Standards for New York State, and he also served on the New York City schools' Chancellor's Math Advisory Panel.

Dr. Posamentier is a leading commentator on educational issues and continues his long-time passion of seeking ways to make mathematics interesting to both teachers, students and the general public — as can be seen from some of his more recent books:

1. *Math Makers: The Lives and Works of 50 Famous Mathematicians* (Prometheus, 2019)
2. *The Mathematics Coach Handbook* (World Scientific, 2019)
3. *The Mathematics of Everyday Life* (Prometheus, 2018)
4. *The Joy of Mathematics: Marvels, Novelties, And Neglected Gems That Are Rarely Taught in Math Class* (Prometheus, 2017)
5. *Strategy Games to Enhance Problem-Solving Ability in Mathematics* (World Scientific, 2017)
6. *The Circle: A Mathematical Exploration Beyond the Line* (Prometheus, 2016)
7. *Problem-Solving Strategies in Mathematics: From Common Approaches to Exemplary Strategies* (World Scientific, 2015)
8. *Effective Techniques to Motivate Mathematics Instruction* (Routledge, 2016)
9. *Numbers: Their Tales, Types and Treasures* (Prometheus, 2015)
10. *Teaching Secondary Mathematics: Techniques and Enrichment Units*, 9th Ed. (Pearson, 2015)
11. *Mathematical Curiosities: A Treasure Trove of Unexpected Entertainments* (Prometheus, 2014)
12. *Geometry: Its Elements and Structure* (Dover, 2014)
13. *Magnificent Mistakes in Mathematics* (Prometheus Books, 2013)
14. *100 Commonly Asked Questions in Math Class: Answers that Promote Mathematical Understanding, Grades 6-12* (Corwin, 2013)
15. *What Successful Math Teachers Do: Grades 6-12* (Corwin 2006, 2013)
16. *The Secrets of Triangles: A Mathematical Journey* (Prometheus Books, 2012)
17. *The Glorious Golden Ratio* (Prometheus Books, 2012)
18. *The Art of Motivating Students for Mathematics Instruction* (McGraw-Hill, 2011)

19. *The Pythagorean Theorem: Its Power and Glory* (Prometheus, 2010)
20. *Mathematical Amazements and Surprises: Fascinating Figures and Noteworthy Numbers* (Prometheus, 2009)
21. *Problem Solving in Mathematics: Grades 3-6: Powerful Strategies to Deepen Understanding* (Corwin, 2009)
22. *Problem-Solving Strategies for Efficient and Elegant Solutions, Grades 6-12* (Corwin, 2008)
23. *The Fabulous Fibonacci Numbers* (Prometheus Books, 2007)
24. *Progress in Mathematics* K-9 textbook series (Sadlier-Oxford, 2006–2009)
25. *What successful Math Teacher Do: Grades K-5* (Corwin 2007)
26. *Exemplary Practices for Secondary Math Teachers* (ASCD, 2007)
27. *101+ Great Ideas to Introduce Key Concepts in Mathematics* (Corwin, 2006)
28. π, *A Biography of the World's Most Mysterious Number* (Prometheus Books, 2004)
29. *Math Wonders: To Inspire Teachers and Students* (ASCD, 2003)
30. *Math Charmers: Tantalizing Tidbits for the Mind* (Prometheus Books, 2003)

Gavrielle Levine has been a mathematics educator throughout her life, culminating in her senior university faculty position preparing teacher candidates at Long Island University to teach mathematics.

As a child, her parents fostered a curiosity about numerical relationships. Her father, an engineer/accountant, instilled an appreciation of numerical patterns while engaging in games and activities that involved score-keeping where one could acquire two or three points at a successful turn. This encouraged mastering counting by two's and three's. He also asked questions that did not have an obvious answer, long before those concepts were introduced in school. One family favorite was "What is infinity?" the answer to which is 'one more than the largest number you can imagine.' Of course, this required his knowledge of higher mathematics concepts. Her mother, a musician, supported her experiential understanding of whole numbers and fractions through teaching her to read music while she learned to play the piano as a young child. Whole notes, quarter notes, eighth notes, and so on gave meaning to fractions, long before they were introduced in school. These understandings and interests were coupled with younger siblings who became willing "students" to be taught the math just learned in elementary school by their sister, the future math teacher, who solidified and clarified her own math concepts while getting practice in communicating these ideas.

As an undergraduate student at Barnard College, Columbia University, mathematics (B.A.) was the obvious major for someone who loved it. This led to a master's degree in mathematics education (M.A.) at Teacher's College, Columbia University, and teaching in the New York City public schools. The discrepancy between her female students' expectation of their math failure and their subsequent success "doing" math aroused her curiosity. She returned to Teacher's College in a Ph.D. program to explore this surprising phenomenon. Her research examined gender differences in mathematics performance in students at different grade levels. Her finding that gender differences emerge during adolescence is widely supported by related research.

With her deeper theoretical knowledge of mathematics education and her experience as a classroom teacher, Dr. Levine taught at colleges in the New York City metropolitan area, including Hunter College (City University of New York) and C.W. Post (Long Island University). Her specialty is preparing teacher candidates at both the undergraduate and graduate levels to teach mathematics. She is currently on the faculty of Long Island University. During her career, she has prepared hundreds of students who are now teachers and school leaders. In turn, through their questions, confusions, and concerns they have guided and expanded her understanding of mathematics education well beyond her childhood early attempts at sibling instruction.

Extending her understanding of mathematics instruction to parents to enhance their support of their children learning math is a natural application for this mathematics educator. It also brings this math educator's career full-circle, since her parents were instrumental in instilling a love of mathematics. Her hope is that future math educators will continue the tradition, both being inspired by their parents and providing inspiration to their children.

Gavrielle Levine lives in Brooklyn with her husband, Alan Palmer, who is also a mathematics educator.

Aaron Lieberman is a tenured faculty member within the Graduate Counseling Psychology program at Long Island University/Brooklyn. Dr. Lieberman has been acknowledged for his teaching activities, within and beyond the University. He has been awarded the 'Newton Award for Teaching Excellence' by his peers, through the Office of the University Vice President for Academic Affairs. Most recently, he was awarded the honor of "Teacher of the Year" through the award-winning "Education Update."

Dr. Lieberman has demonstrated academic success through his scholarship, which focuses on Interprofessional Education (IPE) and Interpersonal Practice (IPP). Dr. Lieberman has worked closely with physical therapists and speech language pathologists in addressing the counseling needs of clients within their practice, as mandated by their professional scope of practice, through presentations and publications. He is the primarily author of book chapters addressing psychosocial and behavioral issues in clients undergoing physical rehabilitation, published by F.A. Davis Company, and sole author of a recent manuscript on addressing behavioral and emotional considerations in the treatment of communication disorders in the American Journal of Speech Language Pathology.

Dr. Lieberman's had been deeply engaged in providing therapeutic services in a variety of clinical settings to individuals, families and couples. In addition to his early work within a Community Mental Health Clinic, a residential treatment center for adolescents with significant emotional issues, and as a private practitioner, Dr. Lieberman primarily worked within the clinical arm of the York City Public School System. His role there, as a member of the Committee on Special Education (CSE), involved reviewing the evaluations and recommendations of school based professionals, as well as in working with families to meet the needs of children within the special education process. Dr. Lieberman also served as the head of the District's Conflict Resolution Team, represented the school district at formal hearings, and provided training for professional staff members on the provision of appropriate services to children and families with special needs.

Dr. Lieberman is a licensed clinical social worker and a licensed mental health counselor in New York State. He is a graduate of Brooklyn College, CUNY, where he received his Bachelor degree, and a graduate of Yeshiva University for his Master's and Doctoral Degrees. He has been active on the board of a number of philanthropic and community groups, and is a voluntary member of the Medical Reserve Team for the City of New York as a vetted mental health first responder.

Danielle Sauro Virgadamo is a clinical child psychologist whose clinical interests focus on parent training and behavioral problems in children. She received her B.A. in mathematics and psychology from The College of New Jersey in 2010, her M.S. in applied psychology from Long Island University — Post Campus in 2014, and her Psy.D. in clinical psychology from Long Island University — Post Campus in 2016.

Dr. Virgadamo has worked in various therapeutic settings, including private practice, outpatient clinics, inpatient units, and day treatment settings throughout New York and New Jersey. She completed psychology externships at Children's Specialized Hospital in Mountainside, New Jersey and Nassau University Medical Center in East Meadow, New York, where she specialized in work with children and their families. She completed her internship at Astor Services for Children and Families in the Bronx, where she worked with children with disruptive behavior disorders, anxiety disorders, and mood disorders. Dr. Virgadamo completed a two-year postdoctoral position at Cognitive Behavioral Associates, a private practice in Great Neck, New York, where she was foundationally trained in Dialectical Behavioral Therapy (DBT) and worked with adolescents with mood disorders and self-injurious behaviors. She also continued her work with young children and their parents, specifically using Parent-Child Interaction Therapy (PCIT) and other behavioral parent management strategies.

During her training at Long Island University Dr. Virgadamo co-founded Family Check-In, a three-session assessment and referral program for underserved families with children ages 2–7. This program consists of a comprehensive evaluation of parent and child symptoms, a parent-child interaction followed by a reflection of the interaction, goal setting, and a feedback session that includes recommendations and referrals. She served as the Administrative Coordinator for Family Check-In and was responsible for creating clinician treatment manuals, phone screen form, chart checklists, scripts for each session, and feedback forms. Additionally, she was responsible for training first and second year students to implement the program in Long Island University's Psychological Services Center (PSC), the primary training clinic for second-year doctoral students. The program has continued its work with underserved families since its implementation and maintains a significant presence in the PSC. Dr. Virgadamo has presented numerous posters at national conferences regarding the implementation of Family Check-In, and is currently writing an article about the feasibility and acceptability of the program to be submitted for publication.

After graduating from Long Island University, Dr. Virgadamo was invited to serve as adjunct faculty for both the undergraduate psychology department and the clinical psychology doctoral program. At the undergraduate level, she taught Introduction to Psychology and Principles of Psychology and in the doctoral program, she taught Advanced Statistics.

Dr. Virgadamo's research interests focus on school-based mental health interventions, scale development, and mental health in twins. In 2012, she traveled to Kitengesa, Uganda, to assess children, interview caregivers, and collect data to determine the effects of a weekly library group on children's literacy, theory of mind, symbolic play, and school-readiness. In 2013, she coauthored a chapter on school-based mental health programs, and in 2016 she coauthored a second chapter that compared school-based mental health interventions in the United States to those in Australia. In 2016, she developed and validated a scale that measures interpersonal interactions between twins. The measure, called the Sauro Twin Interpersonal Interactions (TwInI) Scale measures levels of dominance/submissiveness, competition/cooperation, and independence/dependence in adult twins.

In clinical practice, Dr. Virgadamo works primarily with children, adolescents, and their families. She specializes in parent training for children with disruptive and oppositional behaviors and additionally works with children and adolescents with anxiety and mood disorders. She has experience with psychological assessment, crisis management, family therapy, and group therapy. Dr. Virgadamo currently lives in Baltimore, Maryland, where she is a clinical child psychologist at Kennedy Krieger Institute specializing in psychological assessment and therapy.

Contents

Introduction

Parents want to support their children in all ways possible. They want to help their children succeed as much as possible, and offer support in many areas to facilitate their children's success. However, some parents do not feel that they are able to provide as high a level or quality of mathematics support for their children as they would like. It is unfortunate that many parents, reflecting common public attitudes, express the opinion that they do not like mathematics, perhaps because they have not had the opportunity to learn mathematics for its usefulness as a life tool or for its beauty and power. Support in the home is essential for children to succeed, since the influence of parents is extraordinarily important in the early years and often into adulthood. Therefore, in order to offset the possibly less-than-ideal conditions for parents to support mathematics instruction in the home, this book provides the missing link so that parents can support their children's learning of mathematics.

Naturally, this could be an uphill battle because many people take pride in admitting that they were weak in mathematics in school and did not like it. Therefore, there is a multiphase challenge to be addressed. First, we would help parents appreciate how much math they actually do know. Then, help them link that to what their children are learning — and why it is being taught the way it is now. This book will show the parents that mathematics can be fun and — in a very conversational fashion — impress them with the joy of mathematics. This will then fortify parents with topics and ideas that could do three things: help their children to understand the mathematics they have been exposed to in school; motivate them to enjoy mathematics because of the fun things that one can do with it; and demonstrate to children the general usefulness of mathematics in everyday life.

In order to achieve these goals, this book addresses in a conversational fashion, parents of school-age children — while remaining sensitive to their possible negatively-preconceived notions about mathematics — to expose them to mathematics in a way that they probably themselves never realized. That is, to provide tools of understanding to parents so that they can confidently work with their children. In other words, enable parents to see the relationship between what they do know and what they need to know in order to support their children. It is hoped that these involved parents will be able to sit with their children and explain concepts that both the parents and their children may not have fully understood, and with the help of this book, parents can provide their children with the clearly worded background and explanations that will not only enrich these youngsters, but also give them a better understanding of what they are actually doing with the concepts introduced in school. This will also enable parents to get the feeling that they have shown their children how they can truly understand mathematics in a more comprehensive fashion. We hope this will deepen everyone's understanding and appreciation for mathematics.

Towards this end, we have developed a very reader-friendly book that will demonstrate to parents that mathematics is not the dense and incomprehensible subject that most people in the general readership think it is: esoteric, distanced, impractical, and certainly not fun. This can be done by 'speaking to the parents' sensitively, allowing them to break from any previous preconceived notions.

Most important, parents must be aware about the powerful connection between their beliefs and their children's beliefs. They often don't realize that their beliefs affect their language and behavior, and children are constantly watching their parents while formulating their own beliefs. As mentioned above, parents need to break away from their previously conceived (often negative) notions about mathematics. When parents have negative views regarding mathematics, those views are taught to their children through their words and behaviors. Parents often speak openly and in front of their children about their disdain for math, and their inability to be successful in math "when they were in school." Children who hear that parents 'gave up' and were not successful, or did not/do not recognize the importance of being successful in math, may internalize this and decide that they do not need to put in the effort to be successful in math. This certainly does not support a growth mindset with respect to learning in general and learning mathematics in particular. Would these same parents share the challenges they faced when learning to read with their children? Would they criticize teachers for introducing new ideas in science or new approaches in social studies? For some reason, it is socially acceptable to brag about one's dislike for math, yet rarely do people boast about their deficits in English or history.

For these and other reasons, it is common to have negative feelings, or to experience anxieties regarding math with greater regularity than say anxiety regarding English or history. These almost universal, and generally negative beliefs about mathematics can raise anxieties about engaging in math. Such reactions can range from mild to significant and can lead to avoidance of mathematics. These feelings and the behavioral reactions they produce have been demonstrated to interfere with math learning, even in their milder forms. Given the importance of math ability for academic, career, and day to day functioning, the ability to recognize negative perceptions, thoughts, and anxieties is an important first step to helping your child overcome such feelings and gain success when engaging with math. A general understanding of mathematics anxiety, at any level, together with some concrete and realistic suggestions to help your child minimize any negative feelings or reactions may help to minimize or remove impediments to real learning and engagement with math. This we address in Chapter 2.

Parents are surprised that the strategies that they learned and used to 'do' math are sometimes different from those taught to their children. Consequently, parents find themselves unable to help their children, despite their children's desire for help and parents' desire to be helpful. In part, parents may not see the relationship between their way of 'doing' math and their children's newly-taught way. Doesn't this underscore the need for conceptual understanding as fostered by the *Common Core Learning Standards*? Once parents understand why they followed a given procedure to solve a math problem or complete a computation, it is likely they would be able to see the relationship between their approach and that of their children.

Parents deserve support for understanding mathematics — something that might have eluded them in earlier years. Once they are made aware of the role they could play at home, they can fortify themselves with mathematical ideas and mathematical topics to which their children are exposed. Some may believe they could never attain this knowledge. We attempt to support parents by approaching the issue in a gentle conversational style: discussing a parent's role in providing for proper student expectation; motivating an interest in the subject so that parents can pass on their interest in the subject; assisting parents to understand basic concepts being taught, so that they can work constructively with their children; and providing entertainment in the subject so that they can vary their work with their children from instructional to fun exposures. Ultimately, parents have to buy into this — or nothing will change. This book provides support for parents to feel they can learn this material and help their children accomplish their goal of mathematics success.

Mindful of the fact that family homes and customs can vary greatly, some suggestions will be made for possible opportunities to carry out these parental responsibilities. Among these would be, sitting with their children for a specific time every day to review daily classroom instruction, providing entertaining dinner table discussions on topics related to mathematics, finding specific mathematical topics that might be related to an activity in which the child might be engaging, or create ways in which mathematics can be shown to be both useful and fun. In short, this section will cover an overview of the parent's responsibility as a mathematical motivator and instructional support beyond the school setting.

In Chapter 3, we provide "demystification" of some of the rules and procedures for arithmetic, which children are exposed to in their normal classroom instruction. Parents may have been told that they should simply memorize and not question these procedures. Understanding the background for some of these rules and algorithms can go a long way in allowing students to better appreciate mathematics, as well as providing a level of "liking" of the subject.

Later on, we provide answers to the questions often asked such as, why we can never divide by zero, but we are hardly ever given a reason why that is forbidden. There are many simple examples that can be used to show how division by zero leads to absurd results, and therefore, is discarded from the realm of mathematical operations. Students are often taught arithmetic algorithms but rarely is it explained why these algorithms actually "work."

These are just two simple examples of the kinds of explanations that could be useful to an elementary school student to help them appreciate why the rules and algorithms that they are told to memorize actually work.

The section on practical applications in everyday life has been carefully developed so as not to show any regional idiosyncrasies, such as urban or rural issues, which might not have common ground. Naturally, the selection of such examples must be age-appropriate and conveyed very clearly to the parents, that is, the readership. One common issue that seems to come up for conversation within a family, especially when preparing a meal, is how much larger a 12-inch pan is than a 10-inch pan. Clearly two inches in their respective diameters would only imply a rather minimal difference of their surface areas. As it turns out, the larger pan is 44% greater than the smaller pan — how could that be? There are lots of such examples in the kitchen, in shopping, in measuring around the house, as well as in family activities that during conversations can bring to life a particular topic that is taught at school. A broad variety of such applications of mathematics in everyday life will be provided in this section. Parents can select those that are most relevant to their child, their culture, and the child's grade level.

There is probably an endless number of simple attractive ideas that can be presented to demonstrate the fun that parents and their children can have with mathematics. Number relationships that are truly unexpected are practically boundless. There are certain properties of some numbers in our base 10 number system, such as numbers 9 and 11, which lend themselves to both entertainment and usefulness. Arithmetic entertainments that can provide practice with algorithms in a fun fashion are included, such as guiding students to use a particular rule that leads into a number loop, demonstrating shortcuts to typical arithmetic processes, or demonstrating paradoxes, each of which carries with it a lesson about the nature of mathematics. This latter example usually impresses students to the extent that they never forget the experience.

Everybody likes an entertaining story, and there are many in the history of mathematics. However, it is important that parents be reminded that they must tell the story in the same fashion that they would tell a friend a joke, by gently leading into it as a story and not merely going right to the mathematical idea to be transmitted. For example, a child, might be asked what book they believe Abraham Lincoln, as a young lawyer, carried in his saddlebags? They will be surprised to hear that he carried a copy of Euclid's *Elements*, since he was fascinated with its collection of logic of the mathematics. Another simple story very often told in the history of mathematics is about young Carl Friedrich Gauss, one of the most famous mathematicians. When he was 10, his teacher wanted to keep his class occupied because he had other things to do and asked the class to add the numbers from 1 to 100. How did young Gauss get the answer in just a few seconds, while the rest of class required an hour or more? (Instead of adding the numbers in numerical order, he decided to add the first and the last to get 101, then the second and next-to-last, to get 101, and then noticed that he would have 50 pairs of 101 to give him a sum of 5,050. Such ingenuity further exhibits the relationships that can make mathematics attractive. There are many other stories in the history of mathematics that are cute and at the same time instructional. We hope to fortify parents with a broad variety of such stories that they can then transmit to their children at an opportune time and in an appropriate tone. Children love to hear stories, and using these to motivate an interest in mathematics is another role parents can play to help popularize mathematics.

Essentially, this book has been written in a highly sensitive fashion to motivate and prepare parents to take an active role in supplementing classroom instruction and thereby supporting their children to gain a better understanding of mathematics while hopefully developing a love for the subject. We need more people in our society to engage in this very important endeavor. Enjoy your journey through this book with a constant "eye" towards how you can use the ideas and concepts to support your child's learning of mathematics while at the same time enriching your child's knowledge of the subject.

Chapter 1

The Role of Parents in Supporting Mathematics Learning

In order to properly learn mathematics, one needs to appreciate it for its power and beauty and not fear it because of the unjust reputation it continues to receive in our culture. Many believe that this is the responsibility of the teacher, namely, to motivate children to like the subject and to make the instruction as interesting as possible. However, the truth of the matter is that the role of the parents is absolutely critical in this endeavor. Parents shape their children's beliefs, attitudes, and behaviors in many areas of their lives, mathematics included. In this chapter we will discuss the role the parent can play to support their child's learning of mathematics.

Why Should I Care About Liking Math?

If you are reading this book, you might have a child who struggles with math or possibly dislikes it. Maybe your child comes home every day and complains about any math related issues he or she is responsible for, such as multiplication, word problems, algebra, geometry, or trigonometry. The words, "When will I ever use this in real life?" are likely part of your child's complaints as he or she sits down to complete his or her "dreaded" math homework. Maybe your child doesn't do his or her homework at all. Why does it even matter? And since you are reading this book, maybe you agree. Math is unfortunately portrayed as frustrating and useless, so why are we torturing our children by making them learn something they believe they will "never use?" We need to dispel these notions and make every effort to model positive attitudes towards mathematics continuously. Let us consider a common scenario that might play itself out in many homes.

Johnny is an 11-year-old boy. Every day when he comes home from school, homework is a battle. His parents attempt to sit down with him and help him with his homework, but he doesn't want to do it, particularly the math homework. Arguments about homework completion are a daily occurrence.

"This is stupid!" He shouts. His mother responds to him, "I know you don't like math. I don't like it either! But you have to just be quiet and do it!"

"But why? If you don't even like it, why do I have to do it?" he inquires.

"Because math is important," his mother reminds him.

But does Johnny's mother really believe that? And does Johnny believe her? It is possible that the answers are "No."

Many children (and adults) struggle with math and may dislike it. It is easy to see why math has gotten a bad reputation. Johnny's parents tell him all about what school was like for them and what their favorite subjects were when they were his age. Oftentimes, they tell him that they "didn't like" math or that it was difficult for them. When Johnny comes home with barely passing grades in math, they sigh with relief and say, "at least he passed."

Yet, Johnny's parents do not do this for other school subjects. In fact, they might read stories with him every night before he goes to bed and take him to the history and science museums on the weekends. When he receives just-passing grades in English, history, and science, his parents ask, "What happened? This is unacceptable!"

So, do we really blame Johnny for not liking math and thinking that it is useless? Maybe not. At times, it seems useless. Unless you are a construction worker, engineer, chef, doctor, nurse, architect, clothing designer or mathematician, when did you last calculate a circumference, employ a logarithm, or find a derivative? What about finding the length of a hypotenuse of a right triangle or solving for x?

Between managing finances, calculating tips, crafting, cooking meals and decorating the house, we are willing to bet you use math more than you realize. Because the truth is, math *is* important, and the skills of problem solving and critical thinking that are clearly demonstrated in math can help children grow into successful adults and will prove immeasurably useful in forming their general reasoning ability, not to mention the many career options that will open to them in this technological era. By keeping a few concepts in mind, Johnny's parents can help him develop a mindset that appreciates math and problem solving and also show him that math does not have to be scary and frustrating. It can be beautiful, understandable, and fun.

This chapter will discuss how parents can use their language, expectations, and behavioral principles to help children succeed in math and even enjoy it. Chapter 2 will explore how anxiety around math, in diverse degrees, is relatively common and will survey some of the reasons behind this phenomenon as well as some approaches and suggestions for addressing and minimizing such anxiety. Chapter 3 will assist parents in learning basic math concepts as they are used in today's elementary grades, while chapters 4, 5, and 6, respectively, will provide you with some practical applications, some amusing aspects of math, and some entertaining stories that you can share with your child to show that it can come "alive." We hope this will give you a more positive picture of this important subject matter. For now, you'll have to take our word for it that math is useful and can be fun, because your child's achievement and confidence in math *is* important and will contribute to his or her academic and professional success.

The Powerful Role That Parents Play

As a parent, you most likely realize the powerful role that you play in your child's life. Children look to their parents for almost everything: shelter, food, clothing, advice, and believe it or not, attitudes. First, you must be aware of the powerful connection between your beliefs and your children's beliefs. You might not realize how your beliefs affect your language and behaviors. Your children are constantly listening to you and watching you while formulating their own beliefs, very often based on your beliefs.

Let's look at some examples of how children learn about their parents' beliefs without them being explicitly taught. Have you ever noticed a parent who always seemed to need to get his or her way in a situation? The parent might cause a scene or put up a fuss if things do not go exactly as he or she wishes. For example, the parent might bark orders at waiters and waitresses and complain that the service in a restaurant is not good enough. Now imagine how his or her children may behave. It is likely that the children will adopt a similar mindset and may not treat others with respect. The children might get in trouble in school or throw tantrums when they are told, "no" to a request. They might hold the belief that their needs are more important than others — which may be the message being unintentionally, but implicitly, taught by the parent. On the other hand, this can work in a positive direction as well. You might know a parent who is extremely generous, caring, and giving. This parent might be someone who always brings gifts or food when visiting and might be the first one to volunteer to help someone with whatever is needed. It is likely that this parent's children learn how to share and go out of their way to make sure other children feel included and have fun. Refer to the following flowchart that depicts how parents' beliefs often affect their children's beliefs and behavior.

How Parent Beliefs Affect Child Beliefs and Behavior

Parent's Belief

Parent's Words and Actions

Child's Belief

Child's Words and Actions

Some of you may not be entirely convinced. It seems at times that children act in a way opposite to our beliefs. For example, how often do your children use poor manners when you want them to be polite, or stay out late when you want them to come home at a reasonable hour? However, children are surprisingly perceptive to the belief systems of their parents and will *generally* act in similar ways. Of course, as children mature, they develop their own beliefs, attitudes, and ways of interacting with others. They also realize that they are independent humans, who can behave in ways that do not need to align with the beliefs of others. They test limits and push boundaries.

However, parents are some of the most influential figures in their children's lives, and they fall into a trap when they *expect* their children to act in a way that is different from the way they act. Think of the parent who always needs to get his or her way. Does that parent *want* his or her child to act the same way? Most likely, the parent does not. Behavioral change can be a complex process that includes a variety of interventions; however, to start the process, parents must reflect on their own behavior and attitudes.

What Does a Child's Attitude Have to do With Math?

You may be asking yourself how belief systems and attitudes are related to being good at math. As we pointed out previously, children develop belief systems and attitudes similar to that of their parents. If you don't like math, chances are, your child will not like math either. While it is possible that genetics may play a role in one's dislike for math, as a parent, your beliefs and attitudes regarding math also greatly affect your child's beliefs and attitudes.

When parents have negative opinions about mathematics, those opinions are transmitted to their children through the parents' words and behaviors. For some reason, it is socially acceptable to brag about one's dislike for math, yet rarely, if ever, do people boast about how much they despised English or history. Because of this social norm, parents often speak openly, and in front of their children, about how much they disliked math! This sends a message to children: math isn't fun; and most people don't like it, so you shouldn't have to like it, either.

To illustrate this point, imagine the following conversation between a mother and her elementary-school-aged child:

Sally: Mom, I need help with my homework. Can you come help me?
Mom: Sure, sweetheart. What do you have tonight?
Sally: Spelling, math, and geography.
Mom: Okay, let's do the spelling and geography together. When your father gets home, he can help you with the math.
Sally: Why can't you help me with the math as well?
Mom: Oh, Daddy's just much better at it than I am. I don't really remember all that stuff. It's been a long time since I learned it.

This conversation seems innocuous. After all, there is nothing wrong with admitting that you don't remember something. However, in this example, Sally is getting the message that math is not important enough for her mother to remember, and Mom obviously does not use it very much. It is highlighted by the fact that Mom is willing to help Sally with her other subjects. It's entirely possible that Mom does not remember all of the state capitals or the countries in Europe; however, she is willing to help Sally with her geography homework anyway, because it does not seem as intimidating. In the future, Sally might brush off math as unimportant or useless when that couldn't be further from the truth. Sally might also perceive that her mother's opinion about her math work is rather low and unessential. This can reduce her motivation to learn or appreciate the subject — something we hope to change in this book.

Another message we inadvertently send to children is how difficult math can seem. Parents frequently mention their inability to be successful in math "when I was in school." Children who hear that their parents "gave up with math" and were not successful at learning the subject or do not recognize the importance of being successful in math may internalize this and decide that they do not need to put in any effort to be successful in math as it is wasted time.

Now, imagine this situation: a child comes home with two test papers — one in mathematics and one in English — each of which is marked with a 70% grade. Imagine what a child thinks, when a parent responds with extreme displeasure that the child got only a 70% grade in English. The parent reproaches the children about their lack of competence in English and how shameful it is to get such a low grade. On the other hand, the parent shows relief that the child passed the math test — albeit with a rather low grade, but nonetheless, passed. Imagine what the child is guided to believe when the parent says, "I am so pleased that you passed your math test, since I couldn't do better myself." Now compare the levels of expectation that the parent has transmitted to the child: you must improve your English competence, while I am satisfied with your competence in math. Hence, the child may react by focusing on the English instruction and neglecting the math instruction.

These kinds of conversations occur frequently and have more of an effect than you may realize. Instead of promoting a growth mindset to learning mathematics or learning in general, these conversations have the reverse effect. Remember, parents fall into a trap when they want their children to like math and succeed in it, but speak about how much they disliked math and how poorly they did in the subject regularly.

Below is a list of common phrases that parents sometimes say to their children along with how a child might interpret a seemingly innocuous phrase.

What Parents Say	What Children Interpret
"I hated math"	Hating math is normal
"I'm no good at math"	My parents aren't good at math, so I don't have to be good at it, either.
"I don't remember how to do this"	Math is useless. My parents don't ever use it.
"Ask your father/mother"	My parent doesn't like math, so I don't have to like it, either. I can get someone else to do it for me.
"At least you passed"	It isn't important to be good at math. The bare minimum is enough.

These phrases discourage children from being open to learning and enjoying math. We advise that you stay away from such expressions as much as possible! However, changing some of your language is just the beginning of what you can do to change your child's attitude about math. You might be wondering what other changes you can make to ensure that you help your child fulfill his or her potential and become confident in his or her math abilities. Read on as we hope to fortify you with ideas to use to motivate a liking for math.

What Parents Can Do

As mentioned previously, parents need to break away from any previously-conceived negative notions about mathematics. This means possibly changing your own attitude about math. For some, this can be a challenging task! Since you are reading this book, you have already shown an interest in being supportive of your child. This is a good first step. Crucial aspects of changing your attitude will include understanding basic math principles, practical applications, and entertaining aspects of math.

An important next step is to work on understanding basic math concepts as they are taught today. We prepared Chapter 3 to help you refamiliarize yourself with the basic concepts taught in the early grades, and also introduce you to the way the concepts are presented today. It is also advisable to familiarize yourself with your child's math textbook(s). You may be surprised that the methods you were taught when you were in school are no longer used. Because of this difference, parents are often left confused and frustrated and believe that they are unable to help their child despite their child's desire for help and the parents' desire to be helpful. In part, you may not see the relationship between your way of 'doing' math and your child's newly-taught way. This can also be a concern for you if you had your basic education in a country other than the United States, where it is quite likely that you may have been taught arithmetic algorithms differently from those taught in the today's American schools. If you are able to understand why you should follow a given procedure to solve a math problem or complete a computation, it is likely that you will see the relationship between your child's approach and yours, and will be able to help him or her. Brush up on the current ways that children are being taught math and make sure that you understand the concepts!

Next, it is important to show a genuine interest in your child's work. This can be difficult if you never enjoyed math growing up or if you still find it boring. If you disliked math or found it boring, do your best to not convey that belief to your child. This doesn't mean you should lie to him or her. Everyone has subjects that they dislike and those that they enjoy. However, because parents and peers frequently discuss their dislike of math, oftentimes children receive the impression that math is a subject to fear. When you are discussing math with your child, demonstrate interest and enthusiasm. Remind yourself that the more interest you show in math, the more your child will understand the importance of the subject.

Once you brush up on current mathematical concepts and show interest in your child's work, you can start establishing reasonable expectations for your child's math and homework performance. Expectations are a tricky concept and can have a positive or negative connotation depending on how you view them. One aspect of expectations that we want you to keep in mind is that *children tend to live up to our expectations of them.*

When thinking about your overall expectations for your child, think optimistically but also realistically. Remember the point about the strong connection between your own beliefs and attitudes and your child's beliefs and attitudes? If you believe that your child is capable of putting forth an effort and succeeding, he or she will believe that, too. And if your child is confident in his or her abilities, he or she will likely work hard, which will give him or her a greater chance at success. Likewise, if you believe your child will always struggle or be lazy, prepare yourself to see that prediction come true.

When we apply expectations to math and homework performance, keep in mind that "reasonable expectations" mean expectations will vary for each child based on his or her abilities and situational factors. For example, asking a child who struggles with math to complete 100 math problems in a noisy, crowded room while he or she is waiting with his or her parent to pick up a sibling from a sports practice is not a reasonable expectation. It is more reasonable to ask the child to complete his or her work in a quiet room with no distractions or to ask the child to only complete a subset of the problems. Regardless of the specific expectation, be sure to challenge your child at the same time you set him or her up for success.

So how do we decide what is reasonable and what is a challenge? Start by noticing what your child is already doing. Keep in mind your child's natural abilities and the environment that he or she is in. You likely have an idea of what your child is capable of based on his or her past academic performance. Therefore, it makes sense to pay attention to the environment and how your child works best. Perhaps your child works most quickly when he or she is alone, with no noise or distractions. Other children might become distracted more when they are alone because there is no one available to redirect them if necessary. Maybe your child prefers to be in a room with someone so that he or she can ask questions or make light conversation if the child wants to take occasional breaks. Or, your child might enjoy having light music playing. Take note of the situations where your child appears to be most successful in completing academic work. Base your expectations on what you know your child is capable of doing and what the situation entails.

You might be wondering what you should do when you are helping your child with his or her homework, and you don't remember or understand the concepts he or she is learning. It can be anxiety-provoking and frustrating to feel unable to help your child. But not understanding your child's work does not mean that you can't help. First, take a deep breath and read the instructions carefully. If you still do not understand it, ask your child if he or she can explain it to you. Being able to teach a concept is a good indicator of how deeply someone understands a concept, and your child can attempt to demonstrate that he or she is proficient in the topic. If your child does not understand the material, attempt to figure it out with your child, which will help your child build problem-solving skills and will model determination and patience.

While it is ideal to minimize parents' negative talk and to stay away from discussing "how bad we are" in math, it is important to not lie or pretend you know how to do something that you do not, in fact, know how to do. Lying conveys to children that it is not okay to ask for help or that it is not okay to not know how to do something. This might lead to your child lying more frequently, experiencing anxiety about asking for help, and developing perfectionistic thoughts and behaviors. Pretending to understand a concept may also lead to an incorrect explanation of it, which would be detrimental to the child's understanding of the topic. Instead, adopt a growth mindset — a growth mindset says, "Let's figure this out together," instead of, "I don't know how to do that."

Once you have addressed your own mindset about math and expectations for your child, it is time to help change your child's behavior from math-avoidant to approaching math with an open and eager mind. The following section on behavioral principles will help you shape your children's behaviors and make math more fun for them in the process.

Behavioral Principles

The first behavioral intervention that we want to establish with children is implementing a routine. Most children perform better academically and feel better when they have a structure in place. However, many children do not have a daily after-school routine, and therefore, their behavior may be erratic and chaotic. For the purposes of this book, we will focus on establishing an after-school homework routine.

Make it a habit to be involved in your child's homework routine (at least in the beginning). It is important that you show a genuine interest in your child's homework and schoolwork. As mentioned previously, showing interest in children's work conveys that it is important and should be attended to. It also is a way to show children positive attention and engage in a conversation. The start of this homework routine should ideally occur at the same time each day. Here is an example of an after-school checklist that you might create for your child to follow daily. Note that some of the sample tasks may be unnecessary for your child to complete. If that's the case, leave them out, and feel free to include your own priorities.

Sarah's After School Checklist

- ✓ Place backpack at table
- ✓ Hang up coat
- ✓ Get changed
- ✓ Eat a snack
- ✓ Take out homework at table or desk
- ✓ Review with mom and/or dad the homework that needs to be completed
- ✓ Complete as much of my homework as I can
- ✓ Ask mom and/or dad questions about anything I don't understand
- ✓ Finish homework
- ✓ Explain the homework to mom and/or dad
- ✓ Have mom and/or dad check my homework
- ✓ Free time!

The routine should be implemented as consistently as possible. Ideally, it would begin as soon as the child arrives home from school, as distractions such as phones, computers, tablets, and television can delay the start of the routine and are difficult for children to pull themselves away from. However, due to after school activities such as clubs and sports, your child's routine might not start until later in the evening. In these instances, it is important to remember your child's limitations. He or she will likely be much more tired later in the evening and may require several breaks. In some cases it might be better to avoid a late evening and arrange your child's homework to begin after a short break following the school day.

Although this may not be part of your child's daily homework routine, a technique that promotes learning and growth is for children to explain their homework and schoolwork to their parents. A child, knowing that they will be required to explain their work to his or her parents, is more likely to do the homework in a fashion that ensures understanding rather than merely completing the assignment so as to move on with the evening's "free-time" activities. This promotes genuine learning, as it requires the child to crystallize thoughts in a fashion that would be intelligible to the listener — in this case the parent.

Now that you have established a routine that works for you and your child, let us talk about how to motivate children to do things that we want them to do using basic principles of behaviorism. Besides changing your attitudes (and hopefully, your child's attitudes) about math, there are several principles that can be useful in affecting behavioral change. When combined with managing attitudes, expectations, and understanding basic mathematical concepts, change can occur quickly and powerfully. We will start with one of the reasons children often learn to dislike math, *classical conditioning*.

Classical conditioning is a form of learning that affects all animals. You might remember the research conducted by the Russian physiologist Ivan Pavlov (1849–1936), where Pavlov trained his dogs to salivate after hearing the ringing of a bell. He did this by pairing the bell with meat powder, which made the dogs naturally salivate. By pairing the bell and meat powder repeatedly, the dogs began to associate the two and would salivate when hearing the bell alone. While the original study was conducted with dogs, it is important to note that classical conditioning affects *all* animals, including humans, in similar ways. When a neutral situation is paired repeatedly with something that naturally elicits positive or negative feelings, we begin to feel either positively or negatively about the situation. For example, when a teacher is warm, caring, and pleasant to interact with, he or she may be associated with positive emotions and thoughts. Every time a child (or parent) sees the teacher, he or she might smile and feel a positive emotion, such as happiness, relaxation, and/or excitement. Conversely, when a teacher is critical and harsh, he or she might be associated with fear, anxiety, and/or hostility.

Classical conditioning relates to math because oftentimes, math is paired with frustration, poor instruction, and boredom. When math is restricted to such an environment *of course* children will not find it fun! If your child's only exposure to math is to attend school and listen to the teacher lecture, then come home and complete homework, he or she is likely to dislike it — unless, of course the teacher is a spectacular presenter and engaging all the students on a regular basis. Think about how this applies to other subjects. Parents often take their children to science, history, and art museums, or go to aquariums and zoos. Additionally, they speak the English language every day, and many parents read to their children before bedtime and encourage children to read books for pleasure. This hardly ever happens with math. When math is rarely paired with enjoyable activities or demonstrated to be a topic that is used in daily life, children do not appreciate the utility, beauty, and amusement that mathematics has to offer. To reverse this negative view of math, we need to pair it with fun activities.

One way in which children will show interest in math is if they believe it applies to their everyday lives and activities that they enjoy. Does your child enjoy cooking, baking, sewing, or knitting? If so, it could be beneficial to teach your child about measurements and apply those concepts to his or her most recent project. Likewise, if your child enjoys sports, teaching him or her about angles, velocity, and the wind resistance of a ball flying through the air are real world applications of math that he or she can find interesting and helpful. Of course, this means first brushing up on those topics yourself!

Another way in which math can be portrayed as entertaining or fun is for parents to foster children's interest in games and puzzles that are math-related. Examples of such games are Sudoku and Rubik's cubes. However, a puzzle or game does not have to be overtly mathematical for it to promote critical thinking and problem-solving skills. For children of any age, riddles, treasure hunts, board games, and card games promote skills that can generalize to solving math problems (for example, handling money in Monopoly or using the process of elimination and critical thinking in the popular board game, Clue). For younger children, playing with constructive toys (blocks, Legos, K'nex, etc) can help with the development of visual-spatial skills, and for adolescents and teenagers, brain teasers and Sudoku can help them learn to think critically, develop planning skills, and solve problems creatively. If your child is already intimidated by or dislikes numbers, try starting him or her off with easier level word puzzles such as word searches and basic crossword puzzles so he or she becomes acclimated to finding solutions to prompts. From there, gradually introduce number puzzles and/or riddles.

Be wary when allowing children to use electronic devices, such as computers, video game consoles, and tablets. Although there are many fun, educational, and interesting games and applications available for children that could foster their development of critical thinking and problem-solving skills, many games are mindless and repetitive. Additionally, video games tend to be addicting and are hard to pull children away from. Furthermore, increased screen time leads to less time for interacting with friends and family members and completing real-life responsibilities.

Encourage dinner table discussions that focus on a broad interest in learning. Dinnertime is a great time for family members to connect about how their days went and to discuss any topics of interest. An example of something that could be included in your dinner routine is for each family member to mention something interesting about his or her day. For children, it can be something that they learned in school that day. Children often find it fun to teach their parents. If your child selects something that he or she learned in school that day and teaches it to you, he or she will demonstrate mastery of the material and enjoy being able to teach you as the parent. Other educational topics could also be discussed, including fun applications of the subjects learned that day. As mentioned previously, chapters four and five will give you more ideas for practical and fun applications of mathematics.

While the applications of classical conditioning discussed earlier are a great way to showcase how much fun math can be, they will not be enough to change behaviors in which children already engage. One way to change children's behavior is to use something called *positive reinforcement*. We use positive reinforcement when we want to get people to do something more frequently or more often. For children, our list of things we want them to do more frequently might include:

- Complete their homework properly and on time
- Work independently
- Put forth a proper effort
- Stay focused on one task at a time

- Ask questions when they need help with something
- Tell us about their day at school

The list can go on and on, and probably also includes behaviors unrelated to schoolwork (for example, getting along with siblings, sharing, exhibiting manners, and so forth). Think about certain behaviors that you wish your child did more frequently. Those are the behaviors that we want to reinforce.

We can help children engage in behaviors that we want them to do by providing a *reinforcer*, otherwise known as a reward, for their behavior. A reinforcer is anything that makes it more likely that someone will engage in a specific behavior again. If you go to your job every day because you get paid for doing so, the *money* that you receive in your paycheck is the reinforcer that makes it more likely that you will continue to attend to your job. If you give your dog a treat for complying when you tell him to "sit," and you notice that he becomes more likely to sit when you say the command, the *treat* is the reinforcer.

Children react to various different types of reinforcers — both intrinsic and extrinsic. (A list of common and effective reinforcers will be provided at the end of this chapter for your reference). The easiest, most economical, and oftentimes most effective reinforcer is *praise*. Examples of general praising statements are, "Thank you," "Great job," "I am proud of you" and "Nice work!" Praise is extremely powerful when delivered correctly.

When praising your child, try to make your praise as immediate and specific as possible. Praise delivered immediately after a good behavior will be more effective than praise given hours or days later. It should also be as specific as you can make it for the following reason. Children (and adults) might not know exactly what they are being praised for if they are just told, "Good job!" They might smile and feel good about themselves and think "Well, I did something right!" However, they are much more likely to understand exactly what they did right if you say, "I am proud of you for completing your homework all by yourself today! You didn't need any help from Mommy!" You can be reasonably sure that if you say the latter, tomorrow Johnny will at least attempt to do his homework by himself since he received such specific positive feedback about it. We call this type of praise *labeled praise*, because it puts a label on what the child did right. The following chart is a depiction of the positive behavior cycle. Notice that when a child feels good about something he or she did, the child will likely do it again.

The Positive Behavior Cycle

Even adults are more responsive to labeled praise. Contemplate how you would feel if your boss said to you, "Hey, you are doing really great work." You'd probably feel pretty great, right? Now, think about if your boss said, "You know, I really appreciate how comprehensive your reports have been. Everything I need to find is in your report, and I can see that you put a lot of thought and effort into it." That feels worlds different! Not only do you know exactly what you are doing right, you will probably be more likely to continue focusing on making those reports as comprehensive as possible in the future.

We are going to use this specific type of praise to help children engage in the behaviors previously listed. Because all children are different, as a parent, you are in the best position to generate effective labeled praises that will suit your child's needs. However, because many parents reading this book might be struggling with similar concerns and problems, we have come up with a list of labeled praises that you might find useful.

Praises for homework completion

- Great job completing your homework on time today!
- I am so proud of you for doing your homework all by yourself!
- I love how focused you are right now.
- You are doing such a great job paying attention to your work, even though there are a lot of distractions here.

Praises for bringing home good grades

- I am so proud of you! You must have worked so hard.
- All of that studying really paid off. You did great work!

Praises for sharing information or asking for help

- That was a great question.
- I can tell you really put a lot of thought into that question.
- I love when you tell me about your day in school.
- Thank you for telling me what happened today, even though I know you aren't happy about it.

These statements are just examples. The most important part of delivering praise is to make sure it sounds genuine to you and your child. If it isn't, you'll see that it won't be as effective and could even be counterproductive. So make your statements with true belief!

You might be saying, "But my child *doesn't* do these things! How am I supposed to praise my son for doing his work independently, when he doesn't do that? Or staying focused? Forget it! Those are things I *want* him to do, but he doesn't." Don't panic — there are a few things we can do! The first thing that we can do is to catch your child doing *something* right, even if it is not the exact behavior we are hoping for. Then, we can use something called *shaping* to gradually help him or her exhibit the behavior that we want to see. Shaping means to praise or reinforce any (easy) step in the right direction toward the behavior that we are looking for. Over time, increase the difficulty until you see the target behavior. Here's an example.

What you want to see happening: You want your child to complete all, or most, of her or his homework independently every day.

What's currently happening: You sit with your child and help him or her through each mathematics problem.

What could you do?: After you explain a problem set, wait for your child to complete a problem or two on his or her own. Then, praise your child for completing those problems independently. Or, you can wait until your child does anything else independently and praise him or her for demonstrating independence. Continue to praise your child for showing independence, and gradually increase the amount of tasks that he or she completes by him or herself.

What you could say: "You did great work by figuring out those two problems by yourself!" Or, "I love how you did that all by yourself. You have been becoming a more mature boy or girl." Then, next time — "Wow, you did four problems by yourself! You are getting so good at working independently! I am proud of you."

What you should *not* say: "See? You understand it. Why can't you just do all of them by yourself?" Or, "I bet you could do them all if you really tried." (We will explain more later about certain statements you should avoid)

Why this works: This helps your child feel good about what he or she did at the moment, not feel badly about what he or she hasn't done yet or isn't able to do. It will motivate your child to want to complete more problems independently. Praising independence in other areas of your child's life will have the same effect. Oftentimes (though not always), a child who needs parents there to help with homework requests help in other areas and praising independence in any of those areas will generalize to other areas of life. This means that a child will realize that completing *any* task independently is a good thing.

There will be times when you may want to use rewards or reinforcers other than praise to change your child's behavior. Another way to use shaping to change behavior is to have your child earn daily rewards for accomplishing a target behavior. However, providing tangible daily rewards is not always financially possible or ideal. In this case, you can make a "good behavior" chart to track your child's progress throughout the week, and then provide a reinforcer at the end of the week. It can be helpful to use stickers to track a target behavior. This is especially effective for children in earlier grades but can be useful even in the teen years to keep track of age-appropriate privileges.

Let's look at an example. Take the situation from above where your goal is for your child to be able to complete all of his or her math homework independently. To start the implementation of a behavior chart, sit down with your child and decide upon a fair reward that could be earned by attempting to work more independently. Once a reward is decided upon, it is time to pick the target behavior of the week that you will be tracking. Keep in mind that you want to use shaping to gradually change behavior. Picking the target behavior of "completing all math homework independently" for the first week will likely set your child up for failure. You might select a more attainable target behavior such as "completing five problems independently each day." Then, every day that your child completes the target behavior, place a sticker on his or her chart. Be sure to place the chart in a high profile place where your child will see it often, which will serve as motivation to keep working toward his or her reward. If at the end of the week, your child has earned a preselected number of stickers, he or she would then receive the reward. See the figure below for an example of how a sticker chart might look. In the sample chart, the child received a star every day that he independently completed homework in each subject. On days where he did not independently complete his homework in a given subject, he received an X.

Ryan's Homework Chart				
	Math	English	Science	Social Studies
Monday	★	★	★	X
Tuesday	X	★	★	★
Wednesday	X	★	X	★
Thursday	★	★	★	★
Friday	★	★	★	★
Goal	15 Stars			
Reward	Sleepover with my friend on Saturday night			

A final note about using positive reinforcement is to ensure that you do your best to reward effort and not results. Remember that whatever behaviors you reward will increase only as long as the behavior is within the child's control. How much effort a child puts forth in a given situation is always within his or her control. However, the results of the effort are *not* always within your child's control. Rewarding solely results (such as receiving an A) could cause your child to feel performance or test anxiety, and will sometimes cause children to avoid difficult tasks for fear of failure. Instead, reward perseverance, effort, eagerness to learn, and determination.

Another technique that can help motivate children to do something that they don't like, or want to do, is called *validation*, which is a way of telling your child that you understand his or her feelings, thoughts, or actions. It does not mean that you necessarily agree with your child, just that you understand him or her. Examples of validating statements include:

- I'm sorry to hear that you had a rough day in school.
- I understand that this subject is really hard for you.
- I know you don't like doing your homework.
- I can tell that you feel really discouraged right now.
- I know that this is taking longer than you would like.

Remember, the goal of validation is conveying that you *understand*. Afterwards, you can follow up with an encouraging statement, such as, "I know you can do it! You'll feel so much better once your homework is done, and then you can relax for the rest of the evening."

Sometimes parents are hesitant to use validation because they think that validating children's frustrations, dislikes, and discouragement will make them less likely to engage in the behaviors that parents want to see. If you tell your child, "I understand that you don't like doing your math homework because it is hard for you," you might think that your child will view it as a free pass to not do his or her homework because, *well, mom understands that I don't like to do it!* It turns out that the opposite is true. When children find it really difficult to complete their homework and we pretend that homework is fun or easy, they feel misunderstood, incompetent, and even more frustrated.

When we validate children, we help them feel understood. Once children feel understood, it is more likely that they will listen to our suggestions and advice.

Now that we have reviewed how to praise and validate children, there are a few statements that parents should try to stay away from. The first is using too much criticism. Of course, it is important to give children feedback and critique their work; however, too much criticism leaves children feeling discouraged and incompetent. When children feel unable to succeed, they often stop trying. When criticism is given, it should be constructive, not a personal attack, and combined with encouragement, praise, and/or validation.

Here's an example of criticism that is effective: "It looks like you made the same error on problems 2, 6, and 11. Please go back to those problems, and see if you can figure out what went wrong. I know you're tired and have been working at this for a while. You're doing a great job staying focused, and after this set you'll be done for the night!" This critique incorporates helpful feedback (perhaps the same error was made on all problems), and provides validation (the child might be tired), labeled praise (the child is doing a great job staying focused) and encouragement (the child is almost done for the night and will be able to relax soon).

A less helpful criticism might look like this: "You're making a lot of careless mistakes today. I don't want to hear your whining — you've been complaining this whole time! It's driving me crazy! If you spent less time whining and more time working, you'd be done by now." One reason this criticism is unhelpful is because it does not tell the child what to do. It tells him what he already did wrong without validating that he or she might be having a hard time understanding or focusing. Parents often think that statements explaining what their child did wrong will motivate the child to do better in the future, while the reverse is usually true. So now, this child feels discouraged, incapable, and guilty for frustrating his or her mother. These feelings make it less likely that he or she will have the energy or motivation to focus properly. And even if the child does complete the work promptly, what might he or she expect the response to be? "Finally!"

That takes us to our next trap — accidentally punishing good behavior in children (while pretending that it is praise). Backhanded praise looks something like this:

- See? I knew you could do it, but you never listen to me.
- Why can't you be this focused [*or calm, or independent, or hardworking*] all the time?
- Thank you for *finally* finishing your homework.
- Good job completing those problems by yourself. It would be really great if you could do all of them by yourself, though.

Backhanded praise takes something that your child did well (for example, completing a few problems independently) and makes him or her feel guilty for not doing it all the time, taking too long, or not listening at other times. Instead of making your child more likely to do the desired action more frequently, it makes him or her *less* likely to do it. It might feel necessary at the moment to let your child know that although his or her homework is complete, things were not done exactly the way you would like them to. However, it is not an effective strategy to *actually change his or her behavior*. It simply does not work.

Are we saying that you should never criticize, yell, or become frustrated with your children? Of course not. Parents often get frustrated, angry, and annoyed by things that their children do. No parent is perfect. Our aim in explaining these behavior-change techniques is to provide a set of guidelines that will help parents effectively help their children academically, foster learning, and create a positive, healthy home environment.

One overarching point to keep in mind is that when you provide constructive criticism to your child, keep the criticism simple and only mention one or two points at a time. When a child is confronted with too many comments at one time he or she may become overwhelmed and may remember none of them.

Conclusion

In conclusion, there are many aspects of parenting to keep in mind when helping your child understand and love math. Our goal is not to overwhelm you but rather to equip you with the tools you need to help your child thrive academically.

The process starts with changing your mindset. This means paying attention to your beliefs, words, and actions. Remember that your child listens carefully to you (even if it may not seem like it) and will pick up on your beliefs and attitudes, which he or she may adopt as their own. Avoid speaking negatively about math, teachers, or school in general. In order to change your beliefs and attitudes, it is important to grasp the information that your child will be learning and establish expectations that your child can approach math with an open mind and that you are confident he or she will do so. Find ways that you can explore and enjoy math and share those with your child.

In addition to changing your attitude, there are several effective behavioral strategies that can help children enjoy and succeed in math. Make math fun by taking your child to museums that incorporate mathematical exhibits or by applying math to his or her hobbies and sports. Use positive reinforcement such as praise or prizes to help motivate him or her to work hard. Validate your child's feelings when he or she is frustrated or feels like giving up while encouraging him or her at the same time to push forward. Remember to stay away from excessive criticism and backhanded compliments.

We hope that the following chapters will increase your confidence in your abilities and prove to you that your child's success in math is possible!

As referenced in the chapter, the following is a list of popular reinforcers, sorted by category:

Non-tangible reinforcers

- Praise
- Affection (hug, kiss)
- Inviting a friend to sleep over
- Having a play date
- Having a special family dinner out
- Selecting which movie you will watch together
- Pushing bedtime back by a half hour
- Extending curfew
- Spending quality time together

- Engaging in a fun activity (craft, sport)
- Driving child to school instead of taking the bus

Tangible Reinforcers

- Special toy that child has been asking for
- Legos or blocks
- Bubbles
- Play-doh or modeling clay
- Cosmetics/make up
- Clothing or shoes
- Manicure/pedicure
- Bicycle, scooter, or skateboard
- Money
- Money toward a big-ticket item (for example: future car, bike)

Electronic reinforcers

- Extra TV time
- Time on iPad or tablet
- Time on game console
- Computer time

Edible reinforcers

- Selecting breakfast, lunch, dinner, or dessert for a day
- Sweets (candy, cookie, ice cream)
- Popcorn, chips, or pretzels
- Fruit
- Child's favorite food

This rather detailed journey through the role of the parent is one that we feel is essential for the success of students. Continue with us now as we go through some of the essential aspects to consider for this important responsibility.

Chapter 2

Mathematics Anxiety

Any discussion about helping a child increase confidence, skill, and ability in mathematics would be incomplete without a serious discussion of math anxiety. Many people have some type of anxiety, and anxiety driven by any variety of situations or activities is quite common. For some of us, it is mathematics and arithmetic which give rise to anxious feelings at potentially various intensities. Regardless of the type, or focus of anxiety, anxiety can interfere with higher mental functioning and impede memory, attention, and reasoning abilities. Such potential interference sits alongside of the powerful physical discomfort, which anxiety can often produce. Math anxiety is no exception, and unless you have experienced it personally, you may not be aware of how pervasive and impactful this can be. Math anxiety is prevalent and affects people of all ages. It is estimated that up to 50% of children have some form of math anxiety, with a higher incidence in girls, particularly after the sixth grade.

Imagine you are driving your child to school. While looking in the rearview mirror, you notice your child appearing distracted and uncomfortable. You may ask, "What's wrong?" and your child may respond, "I don't want to go to school. I don't feel well." Knowing that your child has been studying for a math test, you put two and two together to conclude that your child is worried about the math exam. These internalized thoughts, beliefs, or fears may produce outward behavioral signs that can be seen, for example, when the child clearly appears to be avoiding math homework at all costs. Additionally, because there is a universal belief that math ability is highly correlated with general intelligence, children may be sensitive to how they are perceived if they struggle with math. Further, the child may also judge their degree of success, or lack of success in math with their chances of achieving success in other academic areas, placing more stress than necessary upon themselves. Such ideas can certainly have a broad impact on self-esteem and academic motivation, and can result in avoidant behaviors in any activity involving mathematics. Some of these avoidant behaviors can further add to the anxiety levels and higher levels of felt anxiety may result in physical symptoms, including complaints of nausea, excessive sweating, and even heart palpitations.

Math anxiety may dramatically interfere with a child's ability to manipulate numbers and solve mathematical problems, whether it is in everyday life or in academic situations. The child or adult may experience anxiety by the mere anticipation of a mathematical problem or the thought of entering into any situation or activity that may involve math. A child may make statements such as, "I can't do this," "I'm stupid," or "This is too hard." These kinds of thoughts could have been generated from that child's current situational difficulties with math or even earlier negative experiences around math activities, yet could also have been reinforced or picked up from others in the child's environment. Attitudes about math, such as mentioned above, and even the initial feelings of anxiety about math are often picked up from teachers, parents and peers and reflect common societal perceptions about math. These feelings also reflect the prevalence of math anxiety in the population generally and the negative beliefs and false perceptions about math which seem to be universally held by a broad spectrum of the populace. Although math anxiety and the false perceptions and ideas connected with mathematics can potentially impact any individual, there is strong evidence that there exists large attitudinal difference between males and females that result in higher levels of math anxiety in girls. Simply exploring and being aware of their own perceptions and beliefs can help parents make sure they do not send unintended messages to children.

It is often true that entering a math situation with a negative attitude or feeling will interfere with successful functioning in the situation and add to any negative feeling or perception around mathematics. This is true, whether such feelings were derived from early negative experience(s), or from false perceptions or common beliefs devoid of validity. Perceived gender-differences, faulty beliefs, or low self-expectations, will each contribute to a high level math anxiety, of fear, and of math avoidance. In some situations, the ultimate initial goal may not be to help your child figure out how to solve the math problem but rather to overcome their involuntary anxiety reactions and unnecessary worrying about math.

In a society where mathematical ability and skill is crucial for day-to-day functioning, such a skill set is imperative for personal and career success. Math anxiety can be an issue of significant impact and have serious implications for those affected. Therefore, understanding mathematics anxiety and addressing this issue as soon as possible becomes critical as this anxiety has the potential of becoming a long-term, adverse problem.

General Strategies for Reducing Math Anxiety

Addressing the symptoms of anxiety and as many of the contributing factors as possible is critical to helping your child overcome math anxiety. Interventions that involve supportive discussions with the child may help to shed some light and increased insight about the student's reactions and may additionally convey understanding, support, and encouragement for the child experiencing difficulty and anxiety around math. Intervention begins where the child is. Not unlike other personal struggles, the first step is to offer support and understanding, as well as empathy around his or her feelings and experiences as they pertain to the anxiety associated with math. In doing so, it is important to be supportive, refrain from negativity, and avoid minimizing these feelings. Once children become aware of their fear, they can begin to move forward.

Approaching your child with patience, encouragement and understanding may allow the child to take a greater risk and to make attempts at engaging in math activities, despite the child's possible current feeling that this is an anxiety producing activity. Structuring such a supportive atmosphere when working with your child may allow the student to increase their focus as well as to begin to accept that making mistakes or errors is a normal part of the learning of new and complex skills, and need not actually be an over-exaggerated tragedy with negative implications about the child. Making math fun and a normal regular routine, among other approaches detailed below, can reduce the stress, increase focus, and lead to increased competence in math. Improved competency in math, even small incremental improvements, can increase motivation and help break the cycle of increasing anxiety leading to further negative experiences with math, which, in turn, adds to increasing anxiety levels. Disrupting this potentially escalating cycle and reducing anxiety reactions will increase focus and success, develop comfort and increase competency in mathematics.

Approaching your child with a positive attitude and encouraging them to develop a positive outlook can help alter negative thoughts, feelings and perceptions. Additionally, teaching and encouraging the child to use relaxation and focused mindfulness techniques provide simple and effective tools for decreasing anxiety, increasing focus, and gaining self-control. Deep breathing exercises are also effective in reducing anxiety symptoms, relaxing the body and mind, and are tools that are always available and easy to learn. Concurrent with mindfulness approaches, there are other useful techniques in approaching the child who is experiencing math anxiety. Physical activity may serve a similar purpose and outcome from the above, since physical activity has been shown to increase deep breathing, rendering a similar relaxing and focusing effect as the focused deep breathing exercises mentioned above. A brisk walk outdoors, working out on a stationary bike, or even simply pacing around the room can stimulate deep breathing while also providing muscular-skeletal tension release and focus. Any physical activity, either before or during an anxiety-provoking situation, which does not necessitate significant concentration can be helpful in reducing tension and anxiety, while also increasing focus.

Performing simple and repetitive physical activity while engaging in a mathematics activity can be particularly helpful for a younger individual or for an individual who is easily distracted. Mindless physical activity has been seen to help children displace excess energy, increase focus and attention, and can be a help in reducing any anxiety on the part of the child. There are a number of behavioral, cognitive, and physical approaches that can be employed at home and school that are recognized as being helpful in addressing anxiety. A few of these are presented below and might serve as a good starting point toward decreasing math anxiety in your child.

If the math anxiety is severe and persistent, professional advice and professional counseling may be necessary. Getting clarity about, and addressing the specific feelings, thoughts, anxieties, symptoms, and trigger factors for each individual is helpful. There are, however, a number of universal considerations and approaches that may provide positive changes and which may be attempted for most individuals prior to taking a more intensive approach.

However it is approached, getting help earlier, rather than later, can avoid or mitigate the fear and anxiety in math performance, math comfort, math skill, and negative emotional and behavioral reactions to mathematics. Offering patient understanding, assistance, reassurance and support, along with efforts to make math appear fun and interesting can be very helpful. It may also be helpful to occasionally remind ourselves that success with math can drive success in education and subsequently, success in occupational and life endeavors.

Summary of specific recommended helpful ideas and steps for parents

1. Understand the reasons for your child's anxiety

Identifying and addressing the specific causal factors that stimulate and drive math anxieties in your child helps to direct and fine tune your response and attempts to help reduce their anxiety. It is important to observe the behavior and reactions of your child around math activities to determine their attitudes about math but it is crucial to engage them in supportive discussions about their specific fears, thoughts and underlying reactions to math. These will be unique and will differ for each child, yet understanding the specific thought, fear, or belief, of your individual child will help you to tackle that specific thought or fear to help the child overcome or reduce the anxiety around that idea or feeling about math.

To understand a child's specific fears and thoughts about math, it initially becomes crucial for parents to reflect on, and to become aware of their own attitudes about math and the messages they may unwittingly convey to their child about math. When discussing these issues with your child, it is likely that your child will reflect back some version of your own fears, thoughts, and beliefs about math. Therefore, being aware of your own feelings around math becomes a good starting point to understand the reactions of your child and can allow the parent to monitor the messages they may be sending their child. This would help to minimize any negative messages and maximize intentionally planned positive and encouraging ideas about math rather than unintended negative ideas. Statements such as, "Look how I use math every day and how much it helps me," gives the child the message that math is relevant and that it can actually be successfully accomplished. Conversely, statements, such as, "I can't help you with your homework" or "I never learned math this way" sends the message to the student that math must be very difficult if their parents cannot do it.

In general, understanding some of the broad as well as the specific reasons behind the anxiety, allows the parent to counter and to address these specific issues and offer appropriate support and encouragement. Ideas such as "Math is too difficult and I'll never understand it" or "I'm stupid, because I made mistakes and don't understand math," are examples of self-talk which prevent the child from finding success or from even attempting math. Helping the child counter their specific fears and thoughts first requires active listening to verbalizations from your child. If your child expresses the specific thought or idea that he or she does not need to know math, or that it is not important or useful, a parent can subtly encourage the counter idea that it is useful by identifying a familiar situation in the child's life that can be figured out using arithmetic; for example, figuring out with them how long it might take to save their allowance to buy a desired item. Being able to know when they might obtain this desired object could bring anticipatory excitement to your child, and also demonstrate how useful, and doable arithmetic can be. Regardless of the approach taken, acknowledging feelings, thoughts, and fears allows a person to take action and is always the first step to overcome these.

In addition to the above, other reasons behind a student's difficulty with math may relate to situational or medical matters. It may be related to a vision or hearing issue that does not allow the child to see, hear, or focus on what is being taught. Perhaps there is a taller child sitting in front of your child and thereby, partially blocking the view of the board. Perhaps another child is being fidgety or loud, and distracting your child. Perhaps the pace of teaching is not suitable or appropriate for the learning rate or learning style of your child. Whatever the cause, it ought to be determined, addressed, and corrected once you have obtained a real understanding of possible factors involved in your child's difficulty with math. Supportive discussions with your child and/ or the teacher can provide this understanding and guide an appropriate response. Obtaining teacher feedback or consulting with others can provide direction and help, yet it is strongly recommended that open, supportive, and ongoing interactions and discussions with your child can be significantly helpful and illuminating. Gaining the child's perspective and the related fears, thoughts, and feelings can be crucial in revealing underlying factors and shaping helpful solutions.

2. *Shake off mistakes*

While engaging in math activities, errors will certainly be made as a natural part of the learning process. This is true for any new skill or academic endeavor, although common negative attitudes and perceptions about math seem to provide extra pressures and stress when errors are made in math. Hitting a wrong note when learning how to play piano might generate an impulse to try it again to correct this understandable error in learning a complex skill. However, given all of the common fears and beliefs about mathematics, making a math error seems to generate feelings that make the math error difficult to shrug off. Errors made in math carry weight and implications that are rarely produced when learning other academic subjects. A math error, for example, can reinforce the common idea that math is difficult or even impossible to learn. The commonplace idea that math ability is strongly related to overall intelligence only intensifies the implications of making a math error, since such an error may suggest lower intellectual abilities to everyone around you. Negative ideas about math interact and reinforce each other to add even higher levels of stress to any math activity and increase the negative impact of this undue stress.

Helping the child to put his or her inevitable mistakes into perspective, as a normal part of the learning process can be helpful. Helping the child to gain a positive attitude and understanding that mistakes happen and are opportunities for deeper learning and understanding will serve to reduce this pressure and allow for greater success. Reacting calmly about your own mistakes, providing encouragement of a child's continued efforts, and establishing a supportive atmosphere where errors are an exciting opportunity to find the solution could help. Reinforce the idea that errors are natural and that the best way to improve is by making mistakes, correcting them, and then continuing to practice so as to become better at this skill. It is important to avoid giving negative feedback and instead to provide encouraging statements, e.g., "I see that was hard, how about we try it another way." Learning to shake off mistakes can reduce the perceived pressure and fear of math and can then accelerate the learning of math.

3. Start by learning math's fundamental principles and concepts

The idea that children learn by understanding and incorporating knowledge in incremental steps that build upon already acquired knowledge and understanding is certainly applicable to the learning of math. Understanding the ideas behind the process, rather than uncritically applying a process, allows the child to succeed in applying this deeper understanding to any broad variation of a mathematics problem and will help the child to generalize and broadly expand their capacity in applying math principles. There may be a place in math education for rote memorization, especially in the early school years where foundational material is presented, and the language of math must be incorporated for future success and usage. Math is a science and is an academic discipline of ideas and concepts. It is clear that once a student moves beyond the early phases of learning math, deeper success and additional advances in math requires a genuine understanding of mathematics ideas so that these can be applied to a wide range of math problems which can present themselves in very different forms. With a real understanding of math ideas, your child will be given the ability to figure out a solution regardless of how a math problem is presented, instead of being limited to the narrow ability to apply the particular memorized process to those problems that present identically to the problem memorized. This ability to understand math ideas and to then appropriately and successfully apply these concepts, can make doing math an activity to engage with. There is little creativity in applying a memorized process to a given math problem. Making math a creative and engaging experience can make it exciting and further encourage continued learning of math.

One way parents can help their child establish a strong math foundation for future math success, is to ensure that their child learns and understands the ideas and constructs that give shape to the processes and procedures that he or she is taught to apply. The "why" of math can be just as, if not more important, than the "how" of math in ensuring cumulative success and utility in mathematical application. Understanding the principles and concepts underneath a procedure can increase success in math regardless of how it is presented, and is likely to lead to success as the learning progresses and becomes increasingly complicated over the years. Understanding math is a cumulative process of building knowledge and success in incremental steps.

Making learning an active process rather than a rote process can engage children. Make an effort to present math with a positive and excited attitude. Engaging in fun math-related activities can go a long way in establishing positive experiences with math or reversing some of the negative thoughts and apprehensions about math. Adding an understanding of the concepts behind the processes can add excitement and engagement with math at home and at school.

4. Recognize your child's learning style

Individuals learn differently and interact with the world and those around them differently. Some of us may feel that we are able to understand something presented to us best when we hear it when it is presented and listen carefully. Some of us may feel that we need to actually see or visualize a process or a demonstration to fully digest and understand how it is done, while others, may need to write it out many times, to perform, or to physically handle, in order to fully "get it." Trying to determine how your child learns best and presenting and reinforcing the learning by focusing on how that particular child learns best is a recipe for greater success with math. If your child needs tactile cues, find an appropriate relevant activity, such as having the child count marbles, subtract (physically removing) or add marbles to comprehend counting, subtraction or addition. A fidgety, or physically active child, might be asked to move the marbles to different parts of the room, thus being tactile while expending excess energy and making the math activity more fun for a child who can't sit still. Writing out a computational process or a math approach, even multiple times, can help a tactile child to learn and remember a lesson. Repeating the steps out loud, even numerous times as they are applied can help the auditory learner. These are just some examples and suggestions although some experimentation may be helpful to determine what works best for your particular child. Pitch to strength and style, but also recognize that children often learn best through multi-modalities, using auditory, tactile, movement, and visual techniques.

5. Practice

Since math is an academic area requiring a sequential, cumulative, skill set and knowledge, ongoing practice that improves competence and success in math can be rewarding and encourage further engagement. Success builds upon itself and mastering small, easier math skills and knowledge through practice, will set the stage for continued sequential success and increase motivation and competency. Developing a set schedule or routine of math practice can establish a pattern for the student and can begin to make engaging with math a routine, normal part of life. As parents, we also need to recognize the importance of scheduling regular study times as a routine part of the daily life of your child. Cognitive learning theory suggests that attempting the same type of tasks at specifically identified times on a routine basis will psychologically gear the mind and body for greater receptivity and success regarding that activity. Although routinely scheduling math study time consistently at a specified hour at home, along with good study habits, are beneficial, merely engaging in studying math on a recurrent and consistent basis will increase math skill and comfort in and of itself as long as this is done in a supportive and encouraging atmosphere, and at a skill level which is appropriate for the child. This may initially be stressful for the anxious child, but if the skill level of the work allows for some math success, this success can be built upon to encourage further engagement with math going forward.

In addition, using normal daily activities to incorporate and include the child in appropriate math activities can be helpful on a number of levels. Having a child count the change when you are out shopping is an example of taking advantage of common daily math opportunities that can normalize and underline the utility of math, and reinforce or strengthen math skills. In involving the child, it is important to insure that the math task is appropriate to what the child is currently learning, and to ensure that it does not over challenge the current skill set of the child. Such activity should not be presented in a way that might produce anxiety, but rather provide an opportunity to practice a growing skill in a fun way. Making such activity into a game or a fun interaction, perhaps with a reward at the end, increases motivation and positive math attitudes. Counting out the number of coins received as change may be appropriate for one child while calculating the anticipated amount of change which is due, may be appropriate for another, more advanced math learner. There are plenty of daily activities that, with some thought, can be made into a fun interaction around math that can be rewarding because of increased success and because of the interactive involvement with a loving parent or guardian.

6. *Individualize rate of learning*

Whether learning and support come from a tutor, a teacher, or the parent, allowing the student to move at their own learning rate, de-emphasizing the pressure of speed in obtaining solutions, together with providing gentle patience and understanding along the way, assures greater receptivity and a positive learning atmosphere. Every student has their own learning style and learning rate, and although there are a number of factors which can impact learning style and rate, pushing a child or otherwise changing the rate of exposure to math material will lead to increased frustration, anxiety, and failure. Regular and ongoing practice and reinforcement of the math constructs learned will solidify that knowledge and pave the way for ongoing movement in the learning curve. Not practicing, and avoidance of math activities will slow the learning process and, in time, may sabotage advancement. These and other factors will impact not only learning, but also the rate of learning as well, so that ensuring a regular practice schedule, attempting to making math as fun as possible, and setting up a supportive and encouraging learning environment is crucial for positive advancement with math. Knowing the characteristics of the individual child and accounting for them are crucial in maintaining progress and minimizing anxiety around math. Knowledge is best built up incrementally and must be built upon an accurate understanding of the existing level of knowledge and skill. Understanding, teaching to, and reinforcing existing skill and knowledge will help the child feel capable of expanding current skill and will lead to greater understanding and success.

Among the important factors and characteristics of the child are cognitive abilities, concentration ability, and learning rate. Children can only do what they can do, and all efforts to teach them or reinforce their understanding needs to be individualized to their specific needs. Among these, obviously, is the need to continuously gauge the child's progress and adjust the work to meet the needs of that child. Moving too fast, or too slowly, will be frustrating for the child and will only raise discomfort levels and anxieties. Periodic encouragement, reinforcement and tangible rewards for incremental success also contribute to success and lessening of anxiety in math, according to behavioral learning theories.

7. Be patient

Learning mathematics is not dissimilar to the learning of any other academic subject in that it takes time, practice, and patience to learn the codes, symbols, and rules of this as in any subject. It takes practice and repetition to master, knowledgeable and patient teacher(s) to provide direction, a strong support system, and a willing, open, and calm student. Additionally, as noted above, practicing with your child needs to occur at a pace that is comfortable to the learning rate and learning style of the child. This can sometimes cause frustration in the parent who themselves, understands the material being presented, or may perceive the material as being simple or easy, or may be tired of repeating the same idea multiple times. It is important to remember that this is all new information for your child and may be perceived as being difficult. That child may also be anxious, and it is especially important to therefore remember that anxiety will diminish higher cognitive processes and their abilities, making the interaction even more impossible and frustrating for the child. It is not always easy to alter expectations, but allowing your child to move at their pace and repeat processes until they are understood, is the only manner in which the child will master this material. Parents will be well advised to remember that they are there to help the child advance in math and in life, and that their patience is a necessary conduit to these ends.

8. Make it fun

Whatever interests, skills, strengths or weaknesses a child may posses, one thing most children have in common is a desire to engage in fun, interactive games. Many formal games, such as Monopoly, or Monopoly Jr., involve math activities as part of playing the game. The very popular game of Monopoly involves activities such as adding the dots on the dice and then counting the paces forward. The game also involves buying and selling properties, paying and collecting rent, and giving change, which can all be practiced without the usual stress associated with structured math activities.

Playing these games can be a fun way to engage a child and to get a child to practice math skills. Playing math games helps to engage children and helps them to practice skills and to increase their understanding of the relevance of math, yet math games do not have to be limited to formal games or specific circumstances.

Mathematics is so omnipresent that opportunities to engage in fun math activities are all around. In addition to playing games as suggested above, many common hobbies or interests children have can also involve mathematics. Understanding what interests or motivates your child can help you to make mathematics fun and interesting by using that particular interest to engage in math activities. A child interested in sports can track their team's wins and losses, count their sport card collections, or engage in any other math activity specifically related to their interest. Involving the child's specific interest in such activity will maintain motivation for that child.

A parent can engage their children in activities in a way that they do not even realize that they are doing math. Any activity can include counting, adding, subtracting, or other math operations that can be presented as a fun activity or as involving "helping out" the parent, which usually motivates children. This is clearly true when taking your child shopping where there are opportunities to count items purchased, price differences, change received and, so forth. For example, a young child can count out the number of bananas in the bunch, while older children can determine the price per pound of a weighed item. Yet, this is also true for the drive taken to the market, where there are numerals on signs to identify, cars and houses to count or to add up as is appropriate to the level of the child, for example. A parent can involve the child in kitchen activities while cooking or cleaning and can help to count the spoons in the sink or the number of spoons of sugar or other recipe ingredients going into the bowl to happily make cookies. These are simply a few examples of the multitude of daily activities and opportunities to make math fun and to reinforce math learning. These ideas can be helpful for all children but adding such fun math activities to daily activities and interactions can be particularly helpful for children who are naturally math anxious and allow them to practice skills.

Conclusion

Math anxiety is a real and impactful issue that affects children and adults in their academic career and everyday functioning. Understanding the contributory factors that result in a persistent math anxiety is the first step in recognizing and easing negative responses. Addressing the specific feelings and fears of your child and providing support, encouragement, and appropriate resources will help. Teaching and utilizing relaxation and other techniques to reduce the symptoms of anxiety and improve focus, along with behavioral interventions will produce favorable outcomes to reduce math anxiety. Although impactful, math anxiety can be mitigated and math success can be realized.

Chapter 3

Helping Your Child Learn Arithmetic

Mathematics instruction today shares many characteristics with mathematics as you learned it, despite the possibility that it may appear to be different. Adding is still adding, even if the steps your children are being taught may be different from the ways you were taught to add. Perhaps you learned arithmetic in another country, so the steps you learned to solve examples, and the ways you were taught to add, subtract, multiply, and divide, may have been different than those taught in American schools. While the steps you were taught continue to be valid and useful, 'your way' is different from the ways your children are currently taught arithmetic procedures in today's American schools. There are some very real changes in mathematics instruction these days. *Understanding mathematics concepts* is now a major focus of instruction in addition to simply learning strategies to solve problems. Knowledge of computation procedures as well as number facts are important. Models, diagrams, tools as well as electronic resources that are increasingly available to this generation are used to learn mathematics content. Language and terms that might have been considered technical in the recent past are used appropriately and frequently starting in early elementary school. Although this book is largely designed for parents of children of all ages, this chapter will describe some features of current elementary school mathematics instruction and suggest how to make use of them as you support your children in learning mathematics.

Why Mathematics Concepts?

In today's classrooms, your children are introduced to the mathematics concepts supporting computation, as well as to the procedures needed to complete the computation. This may seem unnecessary to an adult who successfully lives a productive life, without being aware of having learned mathematics concepts. It may be that you learned computational procedures and were able to generalize and deduce the concepts on your own. If you learned arithmetic outside of the United States, understanding the concepts supporting arithmetic procedures may bridge the gap between the steps you use to solve a computation and the way your child currently is taught in school. You may be able to see that although your procedures appear different from those of your children, they are in fact equivalent, and they do arrive at or produce the same answer. The advantage for your children of introducing them to the concepts and to the computational procedures is that this approach supports their understanding of *why* problems are solved in a particular way, as well as *when* it is appropriate and effective to use a particular approach. Consequently, children may become more flexible in their choice of solution strategies when solving problems and more creative in approaching new problem situations. Acquiring tools to increase creativity and flexibility may extend beyond the mathematics classroom to enhance other dimensions of your children's lives.

It is helpful for us to consider why children learn mathematics. As with most of their education, children are being prepared to function in the real world. Arithmetic and mathematical knowledge form part of this preparation. In the past, and frequently continuing into the present, computational procedures, also called 'algorithms,' have been taught in a rote manner, modeled by the teacher or textbook, and reproduced by children in response to a well-structured example. The algorithms that parents were taught probably included language like 'carrying' and 'borrowing.' For parents who were taught mathematics in a country other than the United States, chances are the algorithms for arithmetic could have been completely different from those introduced here. New approaches to mathematics instruction focus on the concepts behind the algorithms as well, so that your children will be able to apply their knowledge in the real world, where it counts.

Let's consider an example. In school, your children learn to add. Will they ever be expected to use this knowledge outside of the classroom? Most likely, yes, perhaps on a daily basis: for example, when they purchase more than one item and want to figure out the cost of the total purchase, when they want to know what time it will be in 20 minutes, or when they want to know how many people there are when two classes are combined, they are likely to need to add. As you know, to figure out the answers to these questions, children must be able *to know* that they need to add, and they need to be able *to 'set up' the example*, as well as *compute correctly*. In the real world, examples are not set up for us when we need to figure out the answer. Understanding what it means to add supports your children successfully applying the addition algorithm and solving the problem. We can extend this idea beyond addition, to finding the area of a garden, perhaps with measurements that include fractions. To be prepared to face life with all of its potential mathematical challenges, your children need the skill-set to be able to know what needs to be done, the knowledge to do it, and the confidence to know that they can do it.

How can parents help?

1. The first way that parents can be helpful is to acknowledge and appreciate the value of understanding mathematics concepts and algorithms, as well as the effort involved. Children feel supported when their struggles are recognized, and when they are not expected to meet them alone. You can acknowledge that the task is large, and also that it is important and must be attempted, tried with persistence, and hopefully completed.

2. The more familiar parents are with the mathematics content their children are learning, the more they can effectively support their children. You may want to take a look at your children's textbooks and see what material is presented and how it is done. This may be different from the way you learned it, or currently think about it. Parents who can reproduce the computational procedures their children are taught, then try to understand why the computation takes the form it does, and when to use it, can provide technical assistance to their children. The challenge for parents is to align the form that you were taught with the form your children are learning. We know that the mathematics has not changed (adding is still adding!), but the presentation of how computation is done may appear different. What is most useful for your children is to help them in ways that are consistent with the ways they are being taught, rather than asking them to learn your way first and then figure out what they need to do in their classroom.

3. If the procedure seems familiar and the underlying concept more obscure, you may initiate a conversation with your child exploring

the concept. Honest confusion can sometimes be clarified through conversation, discussion, or at least formulation of responses to the following questions:

a. What is known? For example, I have 2 apples and you have 3 apples.
b. What is not known? For example, how many apples do we have together?
c. What is our plan for how to answer the question? For example, add your apples and my apples.

Mathematics Concepts: Figuring Out 'How Many?'

The over-arching purpose of elementary school mathematics is to answer the question 'How many?' which is presented in a variety of forms. We will consider how the four arithmetic operations answer that question and how to know which operation is best to use in a given situation.

We often think of the arithmetic operations of addition, subtraction, multiplication, and division as each having one unique meaning. In fact, each operation has more than one interpretation that gives meaning to the operation and supports our application of the operation in different situations. Understanding these interpretations increases your children's likelihood of appropriately applying mathematics operations. Each operation serves to answer the question "how many?" in its unique way, consistent with the requirements of the situation, problem, and/or example. A sample of terms associated with each interpretation is in parentheses following the description of the interpretation. Each interpretation is introduced using a diagram to clarify the meaning. Later, notation is introduced to record the relationship and compute a more complex version. A brief overview of the interpretations of arithmetic operations follows. More detail can be found in your child's textbook or online.

Addition

We think of addition as combining two collections of objects to find out how many are in the combined collection. Another term used to describe addition is *composing*, which refers to putting two collections together. There are two interpretations of addition that are closely related, and that your children are likely to learn in their classrooms. One is called 'combining' or 'putting together,' while the other is called 'adding to.' Each interpretation of addition is described separately.

Sometimes two collections are united (*'combined'* or *'put together'*). An example of this interpretation is:

> *You have three apples and your friend has two apples. How many would there be if the apples are combined?*

We can represent each apple with 'O' in a diagram. We count the combined collection to determine how many apples there are.

O O O   O O
your apples   *your friend's apples*
      O O O O O
your apples and your friend's apples combined

This can be represented in an equation as $3 + 2 = 5$.

Another interpretation of addition increases one collection ('**add to**'). An example of this interpretation is:

> *Your basket has five apples and you place four more apples in your basket. How many apples do you now have in your basket?*

We can represent each apple with 'O' in a diagram. We count the collection of apples composed of your apples and those added to your apples to determine how many apples there are now.

O O O O O   O O O O
your basket of five apples   *additional apples added to yours*
     O O O O O O O O O
    your apples with the apples added to yours

This can be represented in an equation as $5 + 4 = 9$.

Subtraction

We subtract to find how many more there are in one collection than in another. This can take several forms. Subtraction is considered *decomposing*, since we start with a whole collection and remove part of it. There are several interpretations of subtraction: 'take away,' 'comparing,' and 'missing addend' (or 'How many more?'). Each interpretation is described separately.

The most familiar interpretation is taking one collection from another (*'take from' or 'take away'*):

> *You have four apples and you take one apple away. How many apples do you have left?*

We can represent each apple with 'O' in a diagram. We count the collection of apples after some were taken from the original collection to see how many remain.

O O O O   *your original collection of apples*
O O O ⊘   *your original collection of apples showing one taken away*
O O O   *your remaining apples*

This can be represented in an equation by $4 - 1 = 3$.

Another interpretation of subtraction is comparing two collections of objects to see which has more (*'comparing'*). An example of that interpretation is:

> *You have four apples and your friend has three apples. Who has more apples? How many more apples?*

This example can be solved by placing both your collection of apples and your friend's collection of apples next to each other, and *matching* apples to see who has more. When no more matches can be made, the collection that still has un-matched apples has more. To find out how many more, we count the un-matched apples. We can represent each apple with 'O' in a diagram.

O O O O   *your apples*
 | | | *matching your apples with your friend's apples*
O O O   *your friend's apples*

After matching your apples and your friend's apples, you have one additional apple that is not matched.

Therefore, you have one more apple than does your friend.
This can be represented in an equation by $4 - 3 = 1$.

A third interpretation of subtraction asks "how many more are needed?"
to complete the collection (***missing addend***). For example,

> *We need 6 eggs to bake a cake, and we now have 2 eggs. How*
> *many more eggs do we need?*

We solve this example by counting the number that is missing: from the
number we have (2) to the number we need (6).
We represent each egg with 'O' in the diagram.

O O O O O O						*number of eggs we need to bake the cake*
O O						*number of eggs we have*
O̶—O̶ O O O O						*number of eggs still needed*
1 *2* *3* *4*						*count the number of eggs still needed*

We can represent this as:

$$2 + \underline{\quad} = 6$$

Four eggs are still needed to bake the cake.
Notice that this example appears to be an addition (there is a $+$ in the
equation). However, we subtract to find the answer.

Multiplication

When we have collections of *equal* size, we can combine them using
multiplication, which is also considered *composing,* since we are combining collections. The 'repeated addition of equal collections' interpretation and the 'array or area' interpretation of multiplication are described
separately.

The most familiar interpretation of multiplication is directly related to
adding collections that are equal in size (***repeated addition of equal collec-***
tions). An example of this interpretation is:

> *You have four baskets of apples. Each basket contains three*
> *apples. How many apples do you have altogether?*

This example can be solved by adding three apples plus three apples plus three apples plus three apples.

It can also be solved by multiplying three apples by four (or four times), since each basket contains the same number of apples. We can represent each apple with 'O' in a diagram.

O O O   O O O   O O O   O O O   *each basket contains three* apples
O O O O O O O O O O O O   *there are 12 apples all together*

This can be represented in an equation by 4 × 3 = 12.

Another interpretation of multiplication is used when objects are presented in a rectangular format (**array** or **area** model). An array is a rectangular distribution of objects, such as a rectangular window with two horizontal panes and four vertical panes or a rectangular chocolate bar that is four sections wide and six sections long. The diagram of an array below shows that side A and side B are opposite each other, and are of equal length to each other.

A

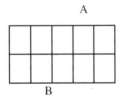

B

Array model

Notice that in the array above, all of the boxes are of equal size. Since they are equal, we can multiply to find out how many boxes there are. We multiply the number of columns (5) by the number of rows (2). We have 10 boxes.

This can be represented in an equation by 5 × 2 = 10.

This interpretation of multiplication is also called the **area model** since it is used to find the area of rectangular regions. To represent the area model, it is best if the interior boxes are squares. Once we have figured out the number of boxes, if they are squares, we know that we have a region of squares, or a region of some number of squares. We can use the area model to figure out this example:

> *We want to cover the floor of a room that is 4 feet wide and 3 feet
> long with a carpet. What is the area of the carpet we need in
> square feet?*

We multiply the number of columns (4) by the number of rows (3), or the
width of the room (4 feet) by the length of the room (3 feet).

We have 12 squares or 12 square feet.

The area of this rectangle is 12 squares or 12 square feet.

This can be represented in an equation by 4 squares × 3 squares =
12 squares or 4 feet × 3 feet = 12 square feet.

Division

When we want to separate a collection into several collections of equal
size we divide, which is considered a *decomposing* operation, since one
collection is separated into equal collections. There are two interpretations of division: 'measurement' and 'distributive.' They are described
separately.

One interpretation of division involves starting with a collection of
objects from which we create several collections of equal size (***measurement***). At the start, we may know how many will be in each new collection
(we know the measure of each new collection) but we do not know how
many collections there will be. For example,

> *If we start with a bag of 10 apples, and want to make smaller
> bags that each contain two apples, how many smaller bags will
> there be?*

We can remove two apples and place them in a small bag. Then we
can remove two more apples and place them in a different bag. We can

continue this procedure until we cannot remove two apples from the original collection. We count the number of bags to know how many smaller collections of apples there are. There are five small bags, each holding two apples.

O O O O O O O O O O   *original collection of 10 apples*
O O   OO   OO   OO   OO   *placing 2 apples in bags*
1   2   3   4   5   *counting the number of bags of apples*

This can be represented in an equation by $10 \div 2 = 5$.

Another interpretation of division starts with a collection of objects and creates a determined number of equal collections (or parts) from it (**distributive**). At the start, we know how many equal collections (how many parts) we want to create. We do not know how many will be in each collection. We distribute the collection equally to the identified parts. For example,

> *If we start with a bag of 10 apples, and want to use them to create two bags with an equal number of apples in each bag, how many apples will there be in each bag?*

O O O O O O O O O O   *original collection of 10 apples*
O O O O O   O O O O O   *placing apples in 2 equal-sized bags*
1 2 3 4 5   1 2 3 4 5   *count each collection*

This can be represented in an equation by $10 \div 2 = 5$.
There are five apples in each of the two bags.

How can parents help?

Learning the concepts associated with arithmetic operations is fundamental for your children to be successful in school. Here are some suggestions for helping your children:

1. Help your children understand the interpretations of the arithmetic operation, what they mean and when they are used.
2. If you can, introduce these interpretations informally to your children, while you are engaged in activities (cooking, shopping, travelling, and

so on), not only when they are doing homework. This will help your children appreciate that knowing these concepts is important in their lives, and in yours.

3. Use tools (discussed below), drawings, and activities to review and explain the concepts.

4. Encourage your children to be the teacher and explain the mathematical operation to you. They are likely to find it difficult to create examples, so that is something you should prepare in advance (from their textbooks, the internet, your creativity).

5. Explain to your children that we need to know *which* operation we need to use before we can use a calculator to do the calculation.

Place Value: 'How Many?'

In order to provide meaning to arithmetic operations, and to utilize this knowledge in our lives, we need a number system that is simple enough to learn, yet complex enough to handle our needs. Ancient systems of numeration included only a few unique numbers, and referred to everything larger as 'more.' Over time, more sophisticated systems included additional symbols, although the highly-regarded Roman system, for example, did not have a unique symbol for zero. The numeration system predominately used today is the Hindu-Arabic system, or Base Ten system. There are 10 unique symbols (0, 1, 2, 3, 4, 5, 6, 7, 8, 9). All other numbers are combinations of these symbols, using specific rules called 'Place Value' which uses multiples of ten to determine their value. For example, a place-value chart may read

Thousands	Hundreds	Tens	Ones

so that when numbers are written in the chart, they derive value by combining their number name and their position in the chart. The number 3579 can be written on the place value chart:

Thousands	Hundreds	Tens	Ones
3	5	7	9

and is read "Three-thousand five-hundred seventy-nine," since the three is in the thousand's column, the five in the hundreds, the seven in the tens, and the nine in the ones. ***Each number is multiplied by the place value***

determined by its position in the number. Only one number between zero and nine may be written in any place value column.

One important characteristic of the place-value model is that every place is ten times the value of the place to its right. For example, the ten's place is ten times one, the hundred's place is ten times ten, the thousand's place is ten times one hundred, and so on.

One consequence of the place-value model is that the number in the position farthest to the left in a number has the largest value in that number. For example, in the number 2,789, the two thousand is larger than any of the other numbers, even though the 2 has a smaller value than the 7, 8, or 9.

The place-value model is considered so important in today's classrooms that your children are probably taught "expanded notation," which, as its name suggests, is a way of writing out the value of each place-value in a number. For example, the number 2,789 can be written in a place value chart to look like:

Thousands	Hundreds	Tens	Ones
2	7	8	9

and can be written in words as 'two thousand seven hundred eighty-nine.'

It can be written using expanded notation as:

$$2789 = 2000 + 700 + 80 + 9$$

or

$$2789 = (2 \times 1000) + (7 \times 100) + (8 \times 10) + (9 \times 1)$$

emphasizing the place-values associated with each numeral. Parentheses are included to clarify which expressions go together.

How can parents help?

You can engage your children in activities emphasizing place value. The basic version of activities is described here; you can develop more elaborate versions to meet the needs of your children.

1. Simulate the base-ten system with tools by 'bundling' ten pencils, straws, or crayons (wrap with a rubber band, string, or ribbon) from

a pile. Creating the bundles will introduce children to the idea of 'tens.' Numbers can be represented by placing a bundle of ten next to three more ($= 13$), and later two bundles ($= 20$), and so on. Ask your children to describe what they did to construct the numbers.

2. Use base-ten blocks (described below) to represent numbers, starting with one through nine. Then, to support children's understanding that *ten ones equal one ten*, ask children to place ten single blocks next to a 10-strip. Since the base-ten blocks are proportional, ten single blocks should be the same length as the 10-strip. Continue asking your children to create numbers to 20, and so on.

3. Play 'Make the largest number you can.' Tools you will need: Either a pair of dice or two sets of index cards with one number from zero through nine written on each card of each set.

   Game play: Ask each child to roll the pair of dice or select two cards and create the largest number possible. Hint: Place the larger number in the ten's column. Ask why this is so. Then place the smaller number in the one's column. Read the number. Can a larger number be made using those two numbers? Repeat with a new roll or card selection. Variation 1: Create the smallest number possible. Ask why this is so. Variation 2: Use three dice or three sets of cards to create the largest/smallest three-digit numbers.

Equal Parts of a Whole (Fractions): How Many?

Not all numbers are whole numbers. Fractions play an important role in our lives. Sometimes they appear mysterious, particularly because of the unusual way they are written.

The most important fact about fractions is that they are *equal parts of a whole*. The number of equal parts into which the whole is divided is called the *denominator*, and the number of parts being considered is called the *numerator*. If we have three slices of pizza from a pizza pie divided into eight equal pieces, three is the numerator and eight is the denominator. In general, a fraction is written like this:

$$\frac{\text{Numerator}}{\text{Denominator}} \quad \text{or} \quad \text{Numerator/Denominator}$$

So, when we write three-eighths, it can look like $\frac{3}{8}$, or sometimes as 3/8.

We can compare fractions to see which is largest or smallest. There are several important ideas to consider:

1. All fractions that are being compared must refer to the same size whole.
2. If the denominators of two fractions are the same, then compare the numerators. The smallest numerator is the smallest fraction. For example, 3/5 is smaller than 4/5.
3. If the numerators are the same, then compare the denominators. *The smallest denominator is the largest fraction.* For example, 2/3 is larger than 2/5. This may seem surprising at first, but remember that the denominator indicates into how many parts the whole is divided. Since it is the same size whole, when dividing into three parts, each part will be larger than when dividing into five parts.

We can extend the concepts underlying the four mathematical operations to fractions. For example, we know that one dozen eggs plus one dozen eggs equals two dozen eggs. But what if we have two containers each of which has one and one-half dozen eggs in it? How many dozen eggs do we have? We add 1½ dozen plus 1½ dozen equals 3 dozen eggs. To figure this out we use the same concept of combining two collections with fractions that we used with whole numbers.

How can parents help?

1. Construct a set of Fraction Pieces as suggested below. Use them with your children to practice comparing fractions.
2. Using Fraction Pieces, find as many ways as you can to show the same size part of a whole (for example, 2/4 = ½).

Mathematical Language

Mathematics is often thought of as the language of science. As such, we use language that may be specific to mathematical situations to describe technical parts of the mathematics experience. Terms may initially be unfamiliar, and may seem artificial and 'too grown up' for children to use. Yet, specific language is used to avoid ambiguity and minimize confusion. Of course, terms must be learned before they can be used, and be helpful. Let's quickly review some arithmetic terms.

In the addition equation $3 + 4 = 7$, both 3 and 4 are *addends*, and 7 is the *sum*.

Were we to use these in the sentence, we would say that instead of referring to 3 and 4 as 'the numbers we are adding together,' we can refer to them as the *addends*.

The answer when adding is called *the sum*. Using the terms addends and sum signals that we are adding. So, we add two addends together to arrive at the sum.

Subtraction terms are not as popular as those for addition, but will be included here for completeness.

In the subtraction equation $7 - 3 = 4$, we refer to 7 as the *minuend* and 3 as the *subtrahend*. In this equation, 4 is the *difference*. When we subtract, we find the difference between the amount we start with (minuend) and the amount we take away (subtrahend).

In the multiplication equation $5 \times 2 = 10$, we often refer to both the 5 and the 2 as *factors*, and 10 as the *product*. Sometimes, 5 is referred to as the *multiplicand* and 2 as the *multiplier*. However, in today's classrooms, *factor* is most frequently used. We create *multiples* of a number when we start with a number, double it, multiply it by 3, then 4, and so on. For example, the multiples of 5 are 5, 10, 15, 20, 25, 30, 35, 40, 45, 50 and so on. Every number has multiples.

In the division equation $10 \div 2 = 5$, we refer to 10 as the *dividend*, 2 as the *divisor*, and 5 as the *quotient*.

One advantage of using specific terms is that they provide information about which operation we are using. For example, when asked to find the quotient, we know we must divide; when told that 8 and 4 are factors, we know that we will multiply; and when told that 3 and 4 are addends, we know that we need to add to find their sum.

When talking about **fractions**, some might refer to the 'top number' and the 'bottom number' of a fraction. Each of those numbers has a name, as we mentioned earlier. In the fraction ½, one is the *numerator*, and two is the *denominator*. These terms provide more information than the location of the numbers. The *denominator* indicates into how many equal parts the whole is divided. In this case, the whole is divided into two equal parts. The *numerator* indicates how many of those equal parts are of interest or are being used in this example. In this case, one of the two equal parts is being used.

How can parents help?

1. When speaking with your children use the mathematical terms when referring to the different roles that numbers play. For example, use 'addends' instead of, or along with, 'numbers we add together.'
2. Encourage your children to use technical terms correctly and appropriately. Help them appreciate that it reduces ambiguity and confusion.

Properties of Arithmetic Operations

Addition and multiplication each have two special properties: The **Commutative Property** and the **Associative Property**.

The **Commutative Property** allows us to add two numbers starting with the first and adding the second, or starting with the second and adding the first. In either case, the answer is the same. For example: $3 + 5 = 8$ and $5 + 3 = 8$.

The Commutative Property is also true for multiplication: $4 \times 2 = 8$ and $2 \times 4 = 8$. We arrive at the same answer, if we start with the first factor and multiply it by the second factor, or if we start with the second factor and multiply it by the first. One way to remember what the Commutative Property does is to think of a 'commuter' who travels the same distance regardless of the whether going from home to work or work to home.

A technical and important part of addition and multiplication is that they are each defined as combining two parts. Each of us has added more than two numbers together or multiplied three numbers (to figure out volume), and it is the **Associative Property** that supports our doing this. When we add three numbers together, we will arrive at the same answer whether we start by adding (*associating*) the first two numbers and then add the third, or we start by adding the last two numbers and then the first. For example, if we want to add 3, 4, and 5, we can add $3 + 4 = 7$, and then add 5, as $7 + 5 = 12$. Or we can add $4 + 5 = 9$ and then add 3, as $9 + 3 = 12$. We often use parentheses to indicate which numbers are added first: $(3 + 4) + 5 = 7 + 5 = 12$ and $3 + (4 + 5) = 3 + 9 = 12$. As you can see, the answer is the same regardless of which numbers we add first, or the notation we use to describe our actions. Similarly, when we multiply three factors, we can multiply the first two together and multiply that product by the third, or multiply the last two factors and multiply that product by the first factor. Regardless of where we start multiplying, the final answer will be the same. For example, if we want to multiply $7 \times 8 \times 9$, we can multiply $(7 \times 8) \times 9$ or $7 \times (8 \times 9)$, since $7 \times 8 = 56$, and $56 \times 9 = 504$ or we can start with $8 \times 9 = 72$ and multiply $72 \times 7 = 504$.

A third property, the **Distributive Property** of multiplication over addition allows us to write numbers as a sum and multiply them in that format. Why? You might ask. Sometimes, writing a number as the sum of the numbers in its place values (write 24 as $20 + 4$) or as the whole number and fraction in a *mixed number* (write 5 ½ as $5 + ½$) simplifies a multiplication or reveals a pattern that otherwise would not be noticed. The key feature of the Distributive Property is that each number in parentheses is multiplied by the same number. That is, we are *distributing* a number over an addition. For example, to multiply

$3 \times 24 =$ ___

we can rewrite 24 as $20 + 4$, and substitute $20 + 4$ in the equation for 24:

$3 \times (20 + 4) =$ ___.

Now, we multiply 3 by each of the numbers in parentheses (*distributing*), and add the result.

So that we get: $3 \times 20 = 60$, and $3 \times 4 = 12$, which by distributing the 3 over the 20 and the 4 gives us: $3 \times (20 + 4) = 60 + 12 = 72$.

How can parents help?

1. Parents can become familiar with the three properties — *commutative, associative, and distributive* — of addition and multiplication and help your children become comfortable with these relationships. They can be helpful tools while your children are solving computations.
2. Identify examples of these properties in 'real life' activities so that children recognize them and value them beyond the classroom. For example, to demonstrate the Commutative Property of addition ask your children to hold 3 pencils in one hand (their left hand) and 4 pencils in their other hand (their right hand). Ask them how many pencils they have. Then, ask them to switch the hands that are holding the pencils so that they are holding 3 pencils in their 'other' hand (their right hand), and 4 pencils in the first hand (their left hand). Ask again how many pencils they

are holding. Now, ask your children to explain how they could have the same the number of pencils, regardless of which hands are holding them. Guide them to relate this to the Commutative Property of addition: The order in which numbers are added does not change the answer.

Mathematical Tools

It is often easier and more meaningful to *see* an example than to hear or say a description of it. As we know, 'A picture is worth a thousand words.' This is also true for mathematical tools that we can move around to see relationships and use to figure out answers. Classroom activities include tools to support children's understanding of mathematical concepts and relationships, as well as provide strategies for children to figure out the answer to mathematics examples. The same or similar tools that are used in the classroom can be used at home. Some mathematical tools will be described here.

Counting pieces: Any small moveable pieces that can be used by children to count, simulate arithmetic operations, or figure out number facts. Round chips made of plastic, cardboard, or wood, as well as M & M's, toothpicks, or pennies are popular (and inexpensive). This versatile tool can be used to represent and figure out all arithmetic operations.

Base Ten blocks: This proportional tool is effective for introducing the place-value model. Small squares (1/2 inch on a side) and 10-strips (5 inches long and ½ inch wide) can be cut from cardboard. Placing 10 squares next to a 10-strip demonstrates that '10 one's is the same as 1 ten.' This tool is also available commercially in plastic and wood. An example of a 10-strip is included here.

Ten-Frames: A rectangle is divided into ten boxes: two rows of five boxes to emphasize pairs of numbers that add to 10. Boxes can be marked to represent each amount, and then counted together.

If ten is involved in an example, the entire Ten-Frame is used, since there are 10 boxes. Ten-Frames can be drawn on paper and counting pieces can be used to mark boxes.

Number Lines: The number line is a straight line with sequential numbers placed at equal distances (similar to a ruler). In elementary school, the number line usually starts at zero. In later grades, negative numbers can be included. Arrows at each end indicate that the line could continue in that direction infinitely.

$$< \quad 0 \quad 1 \quad 2 \quad 3 \quad 4 \quad 5 \quad 6 \quad 7 \quad 8 \quad 9 \quad 10 \quad >$$

Number lines are used with all four of the operations, as well as counting activities.

Fraction Pieces: Strips of cardboard of the same length are divided into equal parts (the denominator) to represent fractions. Frequently chosen denominators are: 1, 2, 3, 4, 5, 6, 8, 10, 12, 20, 24. This tool is ideal to compare fractions visually, so that children see why $1/3$ is larger than $1/4$. A sample of fraction pieces shows the whole, then the whole divided into halves, thirds, fourths, sixths, and twelfths.

[whole strip]	WHOLE
[halves strip]	HALVES
[thirds strip]	THIRDS
[fourths strip]	FOURTHS
[sixths strip]	SIXTHS
[twelfths strip]	TWELFTHS

How can parents help?

1. Construct a set of counting pieces. Encourage your children to use them to model the operations, and to act out examples.

2. Create a set of base-ten blocks. Introduce base-ten blocks to your children. Use them to demonstrate and have your children demonstrate that 10 squares are the same length as one 10-strip. Once this relationship is well-established by your children, ask them to create numbers from one through twenty, first using only the squares, then 'trading' ten squares for a 10-strip.
3. Draw a Ten-Frame on cardboard. Encourage your children to show numbers less than ten by placing counting pieces on boxes. Then, ask them 'How many more are needed to make ten?'
4. Draw a number line on paper or cardboard. Ask young children to use it to count and point to each number as they count. Ask older children to locate numbers without counting, using the number line.
5. Construct 3 identical sets of Fraction Pieces. Leave one set intact as a page-long chart. Cut the next set into strips (of equal length). Cut the third set into individual pieces. Encourage your children to compare fractions by holding pieces next to each other. Ask: Which is larger? smaller? why? More advanced: Practice adding and subtracting fractions with the same denominators using Fraction Pieces. Sample fraction pieces:

	WHOLE
	HALVES
	THIRDS
	FOURTHS
	SIXTHS
	TWELFTHS

Mathematics Strategies: Computing 'How Many?'

Addition

Counting pieces are used to represent the amounts in each example.

To add $3 + 6 =$ ___, set up the example by placing three counting pieces next to six counting pieces. To find the sum, count all of the pieces. The last number counted is the amount in the collection. We can draw a diagram to represent this example, and use 'O' for each counting piece.

O O O   O O O O O O
one collection of pieces   *another collection of pieces*
O O O O O O O O O
the pieces placed together

O O O O O O O O O
1 2 3 4 5 6 7 8 9   *count* all of the pieces

A **number line** can be used to add two numbers.

To add $3 + 6 =$ ___, we locate the first number on the number line by counting from zero moving to the right. For example, we locate 3 on the number line by starting at zero and counting 3 numbers to the right.

$<$ 0   1   2   **3**   4   5   6   7   8   9   10 $>$

start at zero   **stop** *at 3*

Now, to add six to three, we continue counting starting from 3, and count six more.

$<$ 0   1   2   **3**   4   5   6   7   8   **9**   10 $>$

start at 3   **stop** *at 9*

The answer to $3 + 6 =$ ___ is the number where we stopped counting: $3 + 6 = 9$.

Ten-frames can be used to add sums that are less than or equal to 10. For example, we can add $3 + 6 =$ ____ using a Ten-Frame.

To add $3 + 6 =$ ___, we mark the same number of boxes as one of the addends. In this diagram, we marked three boxes using 'O.'

O	O	O		

Now, to complete the example, we mark the same number of boxes as the other addend. In this diagram, we marked six boxes using '×.'

O	O	O	×	×
×	×	×	×	

To find the answer to this example, we count all of the boxes that have been marked.

Nine boxes have been marked.

$3 + 6 = 9$.

The **place-value model** represents each position in a number as a multiple of ten, and is often used with large numbers.

To add $23 + 45 = $ _____, we represent each number in the place-value chart:

```
      10's    1's
       2       3
  +    4       5
```

The first addend, 23 is written as 2 ten's and 3 one's.
The second addend 45 is written as 4 ten's and 5 one's.

Using the place-value model, we combine numbers in the same place value. So, we combine the one's: $3 + 5 = 8$, and the ten's: $2 + 4 = 6$ (actually $20 + 40 = 60$) and write their sums.

```
      10's    1's
       2       3
  +    4       5
       6       8
```

The sum of 23 and 45 is 68: $23 + 45 = 68$.

We know that only a one-digit number can be written in any place-value column. So, what do we do when a sum in a column exceeds nine? We 'regroup' that number as the sum of larger place values, using only numbers between zero and nine in each position. For example, let's add

$23 + 49 = $ ___.

We set up the example the same way as we did above, with addends placed in the place-value chart:

```
   10's   1's
    2      3
+   4      9
```

The first addend, 23 is written as 2 ten's and 3 one's.
The second addend 49 is written as 4 ten's and 9 one's.

Using the place-value model, we combine numbers in the same place-value. So, we combine the one's: $3 + 9 = 12$. Immediately we realize we cannot write 12 in the one's column. We must 'regroup' 12 as tens and ones. We regroup $12 = 10 + 2$ or $12 = (1 \times 10) + (2 \times 1)$ using expanded notation. We write 12 in the place-value chart and create a partial sum:

```
   10's   1's
    2      3
+   4      9
    1      2     partial sum
```

Now we add the tens. 2 tens + 4 tens = 6 tens ($20 + 40 = 60$) and write the partial sum in the place-value chart.

```
   10's   1's
    2      3
+   4      9
    1      2
+   6            partial sum (60)
```

We combine the partial sums to find the answer to this example.

```
   10's   1's
    2      3
+   4      9
    1      2
+   6
    7      2     sum (70 + 2 = 72)
```

$23 + 49 = 72$.

Subtraction

'Take From' or 'Take Away' subtraction

Counting Pieces can be used to figure out 'Take Away' subtraction examples.

To solve:

> *You have 7 apples and you take 5 away to give your friend. How many do you have left?*

We display 7 counting pieces to represent the 7 apples we have. Then, there is an action. From those 7, we take 5 pieces to give to your friend. Now, how many do you have left?

We count the remaining pieces, and know there are 2 pieces or apples left.

We can draw a diagram to represent this example, and use 'O' for each counting piece.

```
O   O   O   O   O   O   O    apples you start with
O   O   O   O   O            apples you gave your friend
                    O   O    apples you have left
```

We write the equation to solve this problem as $7 - 5 = 2$.

We can use a **number line** to figure out 'Take From' subtraction examples.

We start with a number line. We locate the number of apples we start with ($= 7$). This is marked with an $\times$ on our number line.

```
<    0   1   2   3   4   5   6   7   8   9   10    >
─────────────────────────────────────────────────
             O          <------        ×
        ending number   number to friend   starting number
                              5
```

Then, starting at 7, we move to the left the number of spaces representing the number of apples we gave away. Where we stop is the number we still have ($= 2$) which we mark here with 'O.'

The equation for this example is $7 - 5 = 2$.

Ten-Frames can be used to solve 'Take From' subtraction examples.

On a Ten Frame we mark the boxes to represent the number of apples with which we start (= 7). We are using 'O' to represent each apple.

O	O	⊘	⊘	⊘
⊘	⊘			

Then, we cross out the number that we gave away (= 5).

We count the remaining apples that are not crossed out to know how many we still have.

The equation for this example is $7 - 5 = 2$.

We can use the **place-value model** to represent 'Take From' subtraction.

To solve $37 - 25 =$ _____, we start by placing each number on the place-value chart.

```
   10's   1's
    3      7
 -  2      5
```

Starting with the place value furthest to the right, we subtract: $7 - 5 = 2$ and write the answer in the one's column. Then, we move to the column to the left and repeat this procedure: We subtract $3 - 2 = 1$, and write 1 in the ten's column.

```
   10's   1's
    3      7
 -  2      5
    1      2
```

The equation is $37 - 25 = 12$.

What if we need to subtract a number in a place value position that is larger than the one we started with? We need to regroup one ten for ten ones. For example, to solve $37 - 28 =$ ___, we write the example in the Place value model:

```
   10's   1's
    3      7
 -  2      8
```

When we subtract 8 from 7 in the one's column, we see that we need to regroup one ten into ten ones:

10's	1's	we regroup 1 ten for 10 ones	10's	1's
3	7		2	17
− 2	8		− 2	8

We know that we cannot have a number larger than a single digit in any place value position. However, this is temporary, to permit us to subtract.

Now, we can subtract 8 from 17. The equation is $17 - 8 = 9$. We write 9 in the one's column.

We subtract 2 from 2 in the ten's column (really $20 - 20$, since they are 2 tens), and the answer is zero. We can write zero or leave the space blank. The equation is $37 - 28 = 9$.

10's	1's
3	7
− 2	8
	9

Comparison subtraction

We can display comparison subtraction using **Counting Pieces**.

For example, to figure out

> You have seven apples and your friend has five apples. Who has more? How many more?

we display 7 apples, and next to them we display 5 apples. We match each of your friend's apples with one of your apples until one of you runs out of apples. The collection with apples remaining has more.

```
O  O  O  O  O  O  O    your apples
|  |  |  |  |          matching
O  O  O  O  O          your friend's apples
               O  O    you have more apples
```

The equation to represent this is $7 - 5 = 2$.

We can use the **number line** to figure out comparison subtraction.

We start by drawing the number line. We mark one collection of apples above the number line and the other collection below the number line. We match each apple in your line with each apple in your friend's line until there are no more apples to match. The line that still has apples has more.

```
        ×   ×   ×   ×   ×   ×   ×        your apples
<   0   1   2   3   4   5   6   7   8   9   10   >
        O   O   O   O   O              your friend's apples
```

There are 2 unmatched apples in your line. You have 2 more apples than does your friend.

The **place-value model** can be used to solve comparison subtraction examples.

> *If you have 37 apples and your friend has 25 apples, who has more apples? How many more?*

We start by setting up the Place value chart and placing the numbers in it.

10's	1's	
3	7	*your apples*
− 2	5	*your friend's apples*

Then subtract starting with the smallest place value, and continue subtracting each place value column moving to the left.

10's	1's	
3	7	*your apples*
− 2	5	*your friend's apples*
1	2	*you have 12 more apples than does your friend*

Missing addend subtraction

When you are making tens, the **Ten-Frame** is a wonderful tool to use. For example, to solve

> *You need 10 apples to make a pie. You have 4 apples. How many more do you need?*

since the Ten-Frame represents 10, we show the number that we have by marking that number of boxes. Then, we count the unmarked boxes to find out how many more we need.

O	O	O	O	

This Ten-Frame shows we started with four apples of the 10 that are needed. We count the empty boxes to find out how many more we need. There are six empty boxes.

The equation to represent this is $10 - 4 = 6$ or $4 + \_\_\_ = 10$.

A Ten-Frame is often used to *make ten*, which is a useful computational strategy to master. For example, if we have 2 and want to know how many more are needed to make 10, we mark 2 boxes and count those that are not marked.

So, $2 + \_\_\_ = 10$.

O	O			

By counting the unmarked boxes, we see that $2 + 8 = 10$.

Multiplication

Repeated-addition multiplication

Counting Pieces can be used for repeated-addition multiplication.

> *There are 3 apples in a basket and there are 4 baskets. How many apples are there all together?*

We display baskets of 3 apples four times.

O O O   O O O   O O O   O O O   *4 baskets with 3 apples each*

To figure out the total number of apples, we count the apples. Alternatively, we can count by 3's as many times as there are baskets: 3, 6, 9, 12. There are 12 apples all together.

The equation for this multiplication is: $3 \times 4 = 12$.

Place-value model for multiplication

We can use the place-value model to figure out multiplication examples.

We start by writing out the example in a place-value chart. Then, starting with the second factor, we multiply the one's value by each of the values of the first factor, starting with the one's place and moving left, place by place to find the first partial product. We repeat this process with the ten's value of the second factor, this time leaving a space in the one's place before writing the products to find the second partial product. To find the answer, we add all of the partial products together.

For example, to solve 21 × 34 = ____:

10's	1's
2	1
× 3	4

We multiply 4 by 21, starting with the number in the one's position and moving left:

10's	1's	
2	1	
× 3	4	
8	4	*Partial Product from 4 × 1, and 4 × 2 (really 4 × 20)*

Then we multiply 3 by 21, again starting with the number in the one's position and moving left.

Notice that we write the Partial Product starting in the ten's column since we are really multiplying by 30 (=3 × 10).

100's	10's	1's	
	2	1	
× 3	4		
	8	4	
6	3	0	*Partial Product from 3 × 1, and 3 × 2 (really 30 × 20). Of course, the 3 is really 30 (3 × 10). Therefore, 3 × 1 is really 30 × 1, and 3 × 2 is really 30 × 20.*

Notice that we have added another place value column, the 100's, since 30 × 20 = 600.

To find the product, we add the partial products.

100's	10's	1's	
	2	1	
×	3	4	
+	8	4	
6	3	0	
7	1	4	*sum of the partial products*

Notice that since $8 + 3 = 11$, we regrouped 11 as $10 + 1$ and added one more to the 6. Since $600 + 100 = 700$, the final product of 21×34 is 714.

Division

We can represent the *measurement interpretation of division* using **Counting Pieces**. To solve the division:

> *You have 8 apples and want to share them with friends by giving each friend 2 apples. To how many friends can you give apples?*

We begin by displaying the starting number of apples. Then we place them in two's, as shown in the diagram where 'O' represents an apple.

O	O	O	O	O	O	O	O	*starting number of apples*
OO		OO		OO		OO		*apples placed in pairs*
1		2		3		4		*count the pairs of apples*

Now count to see how many pairs of apples there are.

There are 4 pairs of apples, so the apples can be shared with 4 friends.

The equation to represent this division is: $8 \div 4 = 2$.

We can also show the *distributive interpretation of division* using **Counting Pieces**:

> *You have 8 apples and want to share them equally with 2 friends. How many apples can you give each of your 2 friends?*

We start by displaying the starting number of apples. Then we create a place for each of the friends' apples. In this example, there will be 2 places, because there are 2 friends. Now, we distribute systematically (like dealing cards) 1 apple to Friend #1, then 1 apple to Friend #2, then another to

Friend #1, another to Friend #2, and so on until we cannot distribute any more apples to both friends. They must receive the same number of apples. We will display this using 'O' for the apples:

O O O O O O O O   *starting number of apples*
FRIEND #1   FRIEND #2   *places for apples for each friend*
O O O O   O O O O   *equal number of apples to each friend*
4   4   *count the number of apples each friend has*

The equation to represent this division is $8 \div 2 = 4$.

Place-value model for division

We can use the place-value model to compute division. For example, to divide 396 by 3, we write either $396 \div 3 =$ or $3\overline{)396}$ both of which are read 'Three hundred ninety-six divided by three.' Notice that when we read $396 \div 3 =$ we read starting at the left of the equation and continue reading to the right. However, when we read $3\overline{)396}$ we start in the 'middle' of the equation (with the dividend), then read the divisor.

   To divide $396 \div 3 =$ we divide the dividend (396) by the divisor (3). Since 396 can be written using expanded notation as $300 + 90 + 6$, we now divide this by the divisor, which is 3. We start by dividing the largest place value of 396, which in this example is the hundreds. So, $300 \div 3 = 100$. Then we divide the next place value, which is tens, so $90 \div 3 = 30$. Finally, we divide the smallest place value in this example, which is ones, so $6 \div 3 = 2$. Now, we recombine or add the three quotients: $100 + 30 + 2 = 132$. So, $396 \div 3 = 132$.

   When we use the other notation: $3\overline{)396}$ we again divide 300 by 3, and write the answer above the 3, then divide 90 by 3, and write the answer above the 9, and finally we divide the 6 by 3 and write the answer above the 6. The answer is 132 and is written $3\overline{)396}$ with 132 above.

How can parents help?

1. Provide Counting Pieces so that your children can represent the operations they are learning. Start with small numbers and examples that 'work out.' As your children become more competent, increase the size of numbers.

2. Create word problems that can be solved with just one mathematical operation: addition, subtraction, multiplication, or division. Start with small numbers, and gradually increase the size of numbers as your children progress.
3. Provide Base-Ten blocks to develop understanding of the place-value model for the operations. Start with small numbers and examples that 'work out.' Gradually increase the size of the numbers. Help your children see the relationship between the one's and ten's with the blocks and the place-value notation.
4. Focus on the relationship between 1 ten and 10 ones, with both the Base Ten blocks and the place-value model.

Fluency

Your children's classrooms are likely to emphasize procedural and computational fluency. **Procedural fluency** refers to 'automatically' knowing the steps to solving a computation. **Computational Fluency** refers to quickly and accurately knowing number facts.

Procedural fluency is one outcome of thoughtful practice and includes understanding concepts which support correctly knowing the steps to solving a computation. Mindless practice with mistakes is not helpful. Practice doesn't always make perfect, only *perfect* practice makes perfect.

There are several strategies to support computational fluency. Practice is an important component of mastering each of them.

Rounding to the Nearest 10 and Estimating the Answer

Rounding numbers to the nearest ten allows children to focus on the procedure without getting lost in the details of numbers. For example, when adding $43 + 85 =$ ____, children can focus on the addition facts or they can round to the nearest ten and focus on the procedural steps needed to add. Once your children have mastered the procedure, they can return to the example and compute it exactly, if they wish.

How do we round to the nearest ten?

The number line is a helpful tool when children round numbers to the nearest ten. Create a number line, this time from 0 to 100, marking only the tens.

< 0 10 20 30 40 50 60 70 80 90 100 >

To round the first number in the example, ask:

What are the 2 tens this number is between?

For example, 43 is between 40 and 50.

Then ask,

Which ten is closer?

Since 43 is only 3 away from 40, but 7 away from 50, 43 is closer to 40.

40 is the nearest ten to 43.

Let's round 85 to the nearest 10.

Ask:

What are the two tens 85 is between?

We see that 85 is between 80 and 90.

Ask,

Which ten is closer?

Since 85 is the same distance from 80 and from 90, the convention is that when a number ends in a 5, we round up. In this case, we round 85 up to the next 10, which is 90.

Estimating answers

To estimate an answer, round each number to the nearest ten, and compute using those simplified numbers. Then, if you wish, or if it is required, compute the exact answer. For example, estimate $43 + 85 =$ ____.

Rounding to the nearest ten, we can re-write the example as $40 + 90 =$ ____, which is 130.

When we compute the sum using the original numbers the answer is 128. Rather close.

Patterns

There are many patterns in our number system. We can use patterns to support knowing number facts, particularly the multiples of a number. For example, to count by 10's, we have numbers such as 20, 30, 40, 50 etc.

To count by 2's, also called 'skip counting,' say the number 1 softly, then say 2 loudly, then 3 softly, 4 loudly, and so on. The numbers said loudly are the 2's, or the multiples of 2. This technique can also be used to learn the 3's, or the multiples of 3.

The multiples of 9 create a special pattern in our number system. They look like this:

$$9 \times 1 = 9$$
$$9 \times 2 = 18$$
$$9 \times 3 = 27$$
$$9 \times 4 = 36$$
$$9 \times 5 = 45$$
$$9 \times 6 = 54$$
$$9 \times 7 = 63$$
$$9 \times 8 = 72$$
$$9 \times 9 = 81$$
$$9 \times 10 = 90$$

Take a careful look at this chart. What number patterns do you see?

Look at the one's places in the products: They start at 9 for 9×1, and decrease by 1 until we arrive at 9×10, where the one's place is 0.

What about the ten's place? Just the opposite. The ten's place starts with 0 for $9 \times 1 = (0)9$, and increases to 9 at $9 \times 10 = 90$.

Now add the digits in any of the products together: 36: $3 + 6 = 9$, 72: $7 + 2 = 9$. Try them all!

The sum of the digits of a product when 9 is a factor will always be 9.

Here's another pattern. Do you see a relationship between the factor that is multiplying 9 and the ten's place in that product? Look at $9 \times \mathbf{6} = \mathbf{54}$. What about $9 \times \mathbf{3} = 2\mathbf{7}$?

The number in the ten's place is one less than the factor multiplying 9.

How does this help us?

If we want to figure out the product of 9 × 8 = ___, we know that the ten's place will be one less than 8, which is 7. We also know that the sum of the digits is 9, so the one's digit must be 2.

Therefore, without multiplying, we see that 9 × **8** = **72**.

Continue to look for more patterns with the multiples of 9.

Addition and Multiplication Number Facts

Patterns to help your children fluently master addition and multiplication number facts can be organized in Addition Tables and Multiplication Tables.

+	0	1	2	3	4	5	6	7	8	9	10
0	0	1	2	3	4	5	6	7	8	9	10
1	1	2	3	4	5	6	7	8	9	10	11
2	2	3	4	5	6	7	8	9	10	11	12
3	3	4	5	6	7	8	9	10	11	12	13
4	4	5	6	7	8	9	10	11	12	13	14
5	5	6	7	8	9	10	11	12	13	14	15
6	6	7	8	9	10	11	12	13	14	15	16
7	7	8	9	10	11	12	13	14	15	16	17
8	8	9	10	11	12	13	14	15	16	17	18
9	9	10	11	12	13	14	15	16	17	18	19
10	10	11	12	13	14	15	16	17	18	19	20

Addition Table

The Addition Table is a square with the numbers from 0 to 10 along the top and down the left side to form the frame of the table. The numbers inside the frame of the table are sums.

To add 2 numbers, select 1 addend on the top frame of the table and 1 addend on the side frame. For example, to add $3 + 7 =$ ___, find 3 on the top frame and 7 on the side frame. Now, place 1 finger on each number and move the fingers to the center of the table. They meet at the sum of those 2 addends. $7 + 3 = 10$.

+	0	1	2	3	4	5	6	7	8	9	10
0	0	1	2	3	4	5	6	7	8	9	10
1	1	2	3	4	5	6	7	8	9	10	11
2	2	3	4	5	6	7	8	9	10	11	12
3	3	4	5	6	7	8	9	10	11	12	13
4	4	5	6	7	8	9	10	11	12	13	14
5	5	6	7	8	9	10	11	12	13	14	15
6	6	7	8	9	10	11	12	13	14	15	16
7	7	8	9	10	11	12	13	14	15	16	17
8	8	9	10	11	12	13	14	15	16	17	18
9	9	10	11	12	13	14	15	16	17	18	19
10	10	11	12	13	14	15	16	17	18	19	20

Addition Table showing $7 + 3 = 10$

You can point out patterns for your children on this table that will support their computational fluency.

For example, The Identity Element in Addition is zero. That means, any number added to zero is that original number. For example, $8 + 0 = 8, 2 + 0 = 2$. These sums are highlighted on the Addition Table below.

+	0	1	2	3	4	5	6	7	8	9	10
0	0	1	2	3	4	5	6	7	8	9	10
1	1	2	3	4	5	6	7	8	9	10	11
2	2	3	4	5	6	7	8	9	10	11	12
3	3	4	5	6	7	8	9	10	11	12	13
4	4	5	6	7	8	9	10	11	12	13	14
5	5	6	7	8	9	10	11	12	13	14	15
6	6	7	8	9	10	11	12	13	14	15	16
7	7	8	9	10	11	12	13	14	15	16	17
8	8	9	10	11	12	13	14	15	16	17	18
9	9	10	11	12	13	14	15	16	17	18	19
10	10	11	12	13	14	15	16	17	18	19	20

Identity Element in Addition

When a number is added to itself, we call the sum 'doubles.'

These sums create a strong anchor for your children learning addition facts. The Addition Chart below highlights 'doubles.'

+	0	1	2	3	4	5	6	7	8	9	10
0	0	1	2	3	4	5	6	7	8	9	10
1	1	2	3	4	5	6	7	8	9	10	11
2	2	3	4	5	6	7	8	9	10	11	12
3	3	4	5	6	7	8	9	10	11	12	13
4	4	5	6	7	8	9	10	11	12	13	14
5	5	6	7	8	9	10	11	12	13	14	15
6	6	7	8	9	10	11	12	13	14	15	16
7	7	8	9	10	11	12	13	14	15	16	17
8	8	9	10	11	12	13	14	15	16	17	18
9	9	10	11	12	13	14	15	16	17	18	19
10	10	11	12	13	14	15	16	17	18	19	20

Addition Table showing 'doubles'

You can explore the properties of addition using the Addition Table. This table shows the commutative property for addition: $3 + 8 = 11$ and $8 + 3 = 11$.

+	0	1	2	3	4	5	6	7	8	9	10
0	0	1	2	3	4	5	6	7	8	9	10
1	1	2	3	4	5	6	7	8	9	10	11
2	2	3	4	5	6	7	8	9	10	11	12
3	3	4	5	6	7	8	9	10	11	12	13
4	4	5	6	7	8	9	10	11	12	13	14
5	5	6	7	8	9	10	11	12	13	14	15
6	6	7	8	9	10	11	12	13	14	15	16
7	7	8	9	10	11	12	13	14	15	16	17
8	8	9	10	11	12	13	14	15	16	17	18
9	9	10	11	12	13	14	15	16	17	18	19
10	10	11	12	13	14	15	16	17	18	19	20

Commutative Property of Addition

You and your children are likely to find many more patterns in the Addition Table.

Multiplication Table

The Multiplication Table is a square with the numbers from 0 to 10 along the top and down the left side to form the frame of the table. The numbers inside the frame of the table are products.

×	0	1	2	3	4	5	6	7	8	9	10
0	0	0	0	0	0	0	0	0	0	0	0
1	0	1	2	3	4	5	6	7	8	9	10
2	0	2	4	6	8	10	12	14	16	18	20
3	0	3	6	9	12	15	18	21	24	27	30
4	0	4	8	12	16	20	24	28	32	36	40
5	0	5	10	15	20	25	30	35	40	45	50
6	0	6	12	18	24	30	36	42	48	54	60
7	0	7	14	21	28	35	42	49	56	63	70
8	0	8	16	24	32	40	48	56	64	72	80
9	0	9	18	27	36	45	54	63	72	81	90
10	0	10	20	30	40	50	60	70	80	90	100

Multiplication Table

To multiply 2 numbers, select 1 factor on the top frame of the table and 1 factor on the side frame. For example, to multiply $2 \times 6 = $ ___, find 2 on the top frame and 6 on the side frame. Now, place 1 finger on each number and move the fingers to the center of the table. They meet at the product of those 2 factors. $2 \times 6 = 12$.

×	0	1	2	3	4	5	6	7	8	9	10
0	0	0	0	0	0	0	0	0	0	0	0
1	0	1	2	3	4	5	6	7	8	9	10
2	0	2	4	6	8	10	12	14	16	18	20
3	0	3	6	9	12	15	18	21	24	27	30
4	0	4	8	12	16	20	24	28	32	36	40
5	0	5	10	15	20	25	30	35	40	45	50
6	0	6	12	18	24	30	36	42	48	54	60
7	0	7	14	21	28	35	42	49	56	63	70
8	0	8	16	24	32	40	48	56	64	72	80
9	0	9	18	27	36	45	54	63	72	81	90
10	0	10	20	30	40	50	60	70	80	90	100

Multiplication Table showing $2 \times 6 = 12$

Again, you can help your children identify patterns on this table that will support their computational fluency.

For example, the Identity Element in Multiplication is one. Any number multiplied by 1 is itself.

×	0	1	2	3	4	5	6	7	8	9	10
0	0	0	0	0	0	0	0	0	0	0	0
1	0	1	2	3	4	5	6	7	8	9	10
2	0	2	4	6	8	10	12	14	16	18	20
3	0	3	6	9	12	15	18	21	24	27	30
4	0	4	8	12	16	20	24	28	32	36	40
5	0	5	10	15	20	25	30	35	40	45	50
6	0	6	12	18	24	30	36	42	48	54	60
7	0	7	14	21	28	35	42	49	56	63	70
8	0	8	16	24	32	40	48	56	64	72	80
9	0	9	18	27	36	45	54	63	72	81	90
10	0	10	20	30	40	50	60	70	80	90	100

Identity element in Multiplication Table

The product of 2 factors that are the same is called its 'square.' Learning the squares will support your children as they develop computational fluency.

×	0	1	2	3	4	5	6	7	8	9	10
0	0	0	0	0	0	0	0	0	0	0	0
1	0	1	2	3	4	5	6	7	8	9	10
2	0	2	4	6	8	10	12	14	16	18	20
3	0	3	6	9	12	15	18	21	24	27	30
4	0	4	8	12	16	20	24	28	32	36	40
5	0	5	10	15	20	25	30	35	40	45	50
6	0	6	12	18	24	30	36	42	48	54	60
7	0	7	14	21	28	35	42	49	56	63	70
8	0	8	16	24	32	40	48	56	64	72	80
9	0	9	18	27	36	45	54	63	72	81	90
10	0	10	20	30	40	50	60	70	80	90	100

Multiplication Table showing Squared numbers

The Multiplication Table can be used to explore the commutative and associative properties of multiplication. The table below shows the commutative property of multiplication: $2 \times 6 = 12$ and $6 \times 2 = 12$.

×	0	1	2	3	4	5	6	7	8	9	10
0	0	0	0	0	0	0	0	0	0	0	0
1	0	1	2	3	4	5	6	7	8	9	10
2	0	2	4	6	8	10	12	14	16	18	20
3	0	3	6	9	12	15	18	21	24	27	30
4	0	4	8	12	16	20	24	28	32	36	40
5	0	5	10	15	20	25	30	35	40	45	50
6	0	6	12	18	24	30	36	42	48	54	60
7	0	7	14	21	28	35	42	49	56	63	70
8	0	8	16	24	32	40	48	56	64	72	80
9	0	9	18	27	36	45	54	63	72	81	90
10	0	10	20	30	40	50	60	70	80	90	100

Commutative Property of Multiplication

There are many more patterns for your children to discover, which will help them develop fluency.

How can parents help?

In your children's classrooms, developing procedural fluency and computational fluency are likely to be emphasized. You can help your children develop fluency.

1. Spend time with your children looking at the addition table. Identify patterns, and number combinations. Be curious, surprised, and engaged.

Looking at the table does not require memory, but careful looking, which in itself is a valuable skill.

2. Using the addition table as your reference, take turns asking your children questions and answering your children's questions, like "What is 7 plus 8?" The responder can demonstrate on the table how the answer was figured out.

3. Encourage your children to learn the addition facts. Find 'shortcuts' using patterns.

4. Identify patterns and number combinations on the multiplication table.

5. Engage your children with questions that can be answered using the multiplication table.

6. Encourage your children to learn the Multiplication facts. Find 'shortcuts' using patterns.

7. Use the addition table and the multiplication table to explore properties of addition and multiplication: The commutative property and the associative property. These can be helpful as your children develop computational fluency.

8. Introduce your children to the patterns in the multiples of 9. Teach them how to use these patterns to figure out products with 9, without multiplying.

9. Encourage your children to practice skip counting, and learn the multiples of 5, 10, 2, and 3.

10. Eventually, you might encourage your children to memorize the basic arithmetic facts that they will be called on to use spontaneously for the rest of their lives, and that will also make a lot of their future work much easier.

Resources for You and Your Children

Throughout this book you will find lots of resources to support your work with your child. There are many electronic resources available on the internet. Some are of higher quality than others.

Parents' Roles in their Children's Math Learning

This chapter describes current ideas, concepts, and strategies in mathematics education, providing specific suggestions for ways in which parents can help their children. Additional resources are available through your children's textbooks and teachers, on the internet, and through local and national websites describing special features of the mathematics curriculum in your children's classrooms. Parents who are involved in helping their children learn can make a big difference in their children's interest in and understanding of mathematics. Engaging with your children's mathematics learning is another dimension of knowing your children and supporting their development.

Chapter 4

Mathematics at Home and Beyond

Needless to say, there are countless applications of mathematics in the home. Clearly, we cannot cover all of them, and you as parents will need to determine what is appropriate to your home and which examples are inappropriate for your child's experience. Sharing these with your children will help them see the practical applications of math. Doing so will pique their interests and make them more eager and motivated to engage. Let's begin with a problem that could come up in a number of applications. We will use one where a contractor comes into home and wants to install some new duct work. That is, to install a heating or air conditioning passage through an aluminum channel (a duct) which has a rectangular cross-section. (The situation described here actually confronted one of the authors.) It demonstrates that the consideration of area can be extremely critical when considering home construction.

Maximizing Area and Volume

The contractor offered to replace some air-conditioning ducts at a friend's home. Needing to turn the duct around a corner at a right angle in the basement in a rather tight space, he indicated that he would keep the amount of aluminum the same, which was used to make a 6" × 12" duct, but as he would turn the corner, he would need to make it narrower and consequently use a 3" × 15" duct. He claimed that it should be no problem since the same amount of aluminum was being used. We tried to explain to him that this was unacceptable because the cross-section area of the 3" × 15" duct would be considerably less than the 6" × 12" duct, which would have an effect of cutting down the airflow. He didn't understand this point until we chose

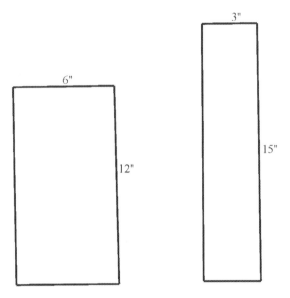

Figure 4.1:   Duct cross-sections

to use as an example of an extreme case, where the dimensions would be 0.5" × 17.5", which would clearly restrict the flow of air, but use the same amount of aluminum. Then he realized the error of his ways.

What we have come to realize with the above example is that it is the *area* of the cross-section of the duct that determines the airflow. In other words, we had to compare the cross-section of a 6" × 12" duct with one that is a 3" × 15" duct. In Figure 4.1 we see the two cross-sections compared, where the first has an area of 6" × 12" = 72 square inches, and the other has an area of 3" × 15" = 45 square inches, which allows less air to pass through than the former.

For the given perimeter of this ductwork, 36", the maximum area would be a square with a side length of 9". Here you have an opportunity to explain to your child that for a given perimeter, the square has the largest area of any rectangle. This is an important concept that you can easily transmit your child.

This can be extended to a related situation, where we have a given amount of cardboard and we would like to create a box with the greatest possible volume from the given amount of cardboard. Rather than to experiment with different dimensions, we can extend the notion of a square providing the largest area from the given perimeter, to the notion that a cube would generate the largest volume. That is, a rectangular solid with all square faces.

Emptying Water from a Pool

Another issue to deal with when it comes to home construction is the problem of emptying a swimming pool or a bathtub, where one has the option of having two drains of 2" diameter, or one drain of a 4" diameter. Before tackling this problem ask your child what his or her intuition dictates. Remember, the way in which this problem is presented to your child makes a big difference in its overall effectiveness. Intuitively, it would seem that either option would drain equally quickly. However, that is not the case. Once again, we need to consider the cross-section of each of the two drain sizes. Using the famous formula for the area of the circle (πr^2) of radius 1, which is $\pi 1^2$, we find the area of the cross-section of each of the two smaller drains is equal to π, which together gives us a total cross-section area of 2π. On the other hand, the cross-section area of the larger drain of radius 2 is 4π, which is twice the drain capability of the two smaller drains. We see that our intuition might have misled us, but mathematics brings us back on the correct path.

Filling a Bathtub

If this counterintuitive example made an impression your child, you might want to take this other step to an analogous problem, which would be that of filling a tub with two hoses. Suppose one hose can fill the tub in 2 hours and the other hose can fill the tub in 3 hours. How long would it take to fill the tub with both hoses working together? Once again, our intuition may not be very helpful here. There are a number of ways of approaching this problem. One is to notice that since the first hose alone can fill the entire tub in two hours, it can fill $\frac{1}{2}$ of the tub in 1 hour. The second hose working alone could fill the entire tub in three hours, it would then fill $\frac{1}{3}$ of the tub in one hour. Working together, they would fill $\frac{1}{n}$th of the tub in one hour. Algebraically, we would represent that as follows: $\frac{1}{2} + \frac{1}{3} = \frac{1}{n}$, and then we have $n = \frac{6}{5} = 1\frac{1}{5}$ hours, or one hour and 12 minutes. Again, we have something that is not intuitively obvious, but if presented in story form, your child would once again have demonstration of how algebra can be used to solve a practical problem at home.

The Perfect Manhole Cover

There are many things in our culture that we seem to take for granted without questioning. Children walk past manhole covers (which in some cultures are referred to sewer covers) and never give a thought to its shape. Taking the time to make children aware of the many shapes they see on a daily basis and and to consider the reason for the shape is an integral part of making mathematics pertinent to our youth.

As we said earlier, considering that a square has the greatest area for a given perimeter could be useful when mapping out a garden area with the given amount of fencing, or any other such areas with a defined boundary. Or when considering a carton, the one in the shape of a cube, where all faces are squares, is the one that would give you the maximum volume for given amount of cardboard. Now let's go back and consider the shape of the manhole cover. Is this relevant here or might there be another reason for their shape?

How many times do we walk over manhole covers without thinking for a moment why they are all of circular shape. The vast majority of manhole covers we see on the street are, in fact, of circular shape. Have you ever wondered why they are all circular? It would be good at this point to have your child speculate about the reason for it being circular in shape. Gradually you should be able to guide your child to the notion that the circular shape cover can never fall into the hole, as would be the case with a square shape cover, as you can see in Figure 4.2.

Now that that has been clearly demonstrated, the question can then be posed, is there any other shape that can be used to cover a manhole that will also not be able to fall into the hole? Here you can now open a new world for your child, something that is clearly not be presented in the traditional school curriculum. But this is the sort of thing that gives you some leverage as a supporter of mathematics instruction in the home and perhaps makes children more curious and motivated for the normal instructional program. The answer to the previously posed question as to there being any other shape besides a circle that can be used as a manhole cover and still not fall in was provided by the German engineer Franz Reuleaux (1829–1905), who

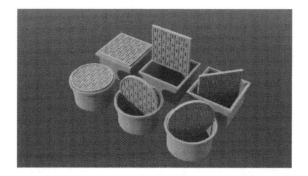

Figure 4.2: Manhole covers

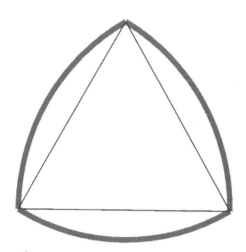

Figure 4.3: Reuleaux Triangle

taught at the Royal Technical University of Berlin, Germany. He developed a rather odd-looking shape that is now called a *Reuleaux triangle*, which we show in Figure 4.3.

One might wonder how Franz Reuleaux ever thought of this triangle. It was said that he was in search of a button that was not round, but still could fit through a buttonhole equally well from any orientation as is the case with a circular button. His "triangle" solved the problem.

The *Reuleaux triangle* is formed by three circular arcs along each side of an equilateral triangle with centers at the opposite vertices of an equilateral

Figure 4.4:   Wrench grabbing a circle

Figure 4.5:   Wrench grabbing a Reuleaux triangle

triangle. It has many unusual properties. It compares nicely to a circle of similar breadth. In the case of a circle, the breadth is the diameter, while for the Reuleaux triangle, it is the distance across — from a triangle vertex to the opposite arc. We refer to the distance between two parallel lines tangent to the curve as the breadth of the curve. In Figure 4.4, we notice how a wrench is simply ineffectual when trying to turn a circular screw.

The same would hold true for a Reuleaux triangular head (Figure 4.5). It, too, would slip, since it is a curve of constant breadth, just as the circle is.

Now getting back to our original manhole cover problem, we can see that the Reuleaux triangle would also satisfy all the conditions of a circle and therefore we can see a manhole cover of the shape as is shown in Figure 4.6.

This becomes particularly significant when fire hydrants are designed so that only special tools can turn them on. Often, a pentagonal valve screw is used. At this point it would probably be clever to ask your child why a pentagonal shape is often used for the valve of a hydrant. They should be guided to realize that unlike a square or a hexagon, a pentagonal shape does not have two parallel sides for a wrench to properly grab onto. Similarly, a circular valve screw would not suffice as a wrench would also not be able to

Figure 4.6: A manhole cover in the shape of a Reuleaux triangle

grab it. This brings us to the point of using a Reuleaux triangle type valve screw, which can only be turned with a wrench of exactly the same shape. One such example of a fire hydrant is shown in Figure 4.7.

So here is a practical application of this situation. During the summer months, kids in a city like to "illegally" turn on the fire hydrants to cool off on very hot days. Since the valve of the hydrant is also often a hexagonal shaped nut, they simply get a wrench to open the hydrant. If that nut were the shape of a Reuleaux triangle, then the wrench would slip along the curve just as it would along a circle. However, with the Reuleaux triangle nut, unlike a circular-shaped nut, we could have a special wrench with a congruent Reuleaux triangle shape that would fit about the nut and not slip. This would not be possible with a circular nut. Thus, the Fire Department would be equipped with a special Reuleaux wrench to open the hydrant in cases of fire, yet the Reuleaux triangle could protect against playful water opening, and avoid water being wasted in this manner.

Figure 4.7:   A fire hydrant nut in the shape of a Reuleaux triangle

In case you have a gifted child, or simply one that would like to pursue this issue further we offer the following discussion of this special "triangle." A curious property of the Reuleaux triangle is that the ratio of its perimeter to its breadth, which is $\left(\frac{\frac{1}{2}\cdot 2\pi r}{r} = \pi\right)$, is the same as for a circle, where the ratio of its circumference to its diameter is $\frac{2\pi r}{2r} = \pi$.

The comparison of the areas of these two shapes is quite another thing, and could be useful when deciding what shape to make a manhole cover. Let's compare their areas. We can get the area of the Reuleaux triangle in a clever way, by adding the three circle sectors that overlap in the equilateral triangle and then deducting the pieces that overlap, so that this triangular region is actually only counted once and not three times.

The total area of the three overlapping circle sectors, where each is $\frac{1}{6}$ of the area of the circle, is equal to $3\left(\frac{1}{6}\right)(\pi r^2)$. From this we need to subtract twice the area of the equilateral triangle, which is $\frac{r^2\sqrt{3}}{4}$. Therefore, the area of the Reuleaux triangle is equal to $3\left(\frac{1}{6}\right)(\pi r^2) - 2\left(\frac{r^2\sqrt{3}}{4}\right) = \frac{r^2}{2}(\pi - \sqrt{3}) \approx r^2\left(\frac{3.1416-1.732}{2}\right) = 0.7048 \cdot r^2$, while the area of a circle with diameter of length r is equal to $\pi\left(\frac{r}{2}\right)^2 = \frac{\pi r^2}{4} = 0.7854 \cdot r^2$.

Therefore, the area of the Reuleaux triangle is less than the area of the circle, which we can also see rather clearly in Figure 4.8.

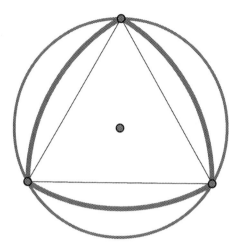

Figure 4.8: Reuleaux Triangle

This is consistent with our understanding of regular polygons, where the circle has the largest area for a given diameter. Furthermore, the Austrian mathematician Wilhelm Blaschke (1885–1962) proved that given any number of such figures of equal breadth, the Reuleaux triangle will always possess the smallest area, and the circle will have the greatest area.

Therefore, from a practical standpoint, to design a manhole cover of a given breadth, and one that would not be able to fall into the hole, the Reuleaux triangle shape would be the economic choice as it would require less metal to construct it. Here is an interesting application of some genuine mathematics applied to a piece of our environment that we seem to take for granted.

By the way, if you feel ambitious enough to take this even one step further, we can also create a *Reuleaux pentagon* by constructing circular arcs from each vertex of a Pentagon as shown in Figure 4.9. This shape shares the properties of the Reuleaux triangle and the circle, as it can fit between two fixed parallel lines regardless of its position. Can this be done with other regular polygons? If so, which ones? This is a good test to see if your child has understood the notion presented about shapes of constant breadth.

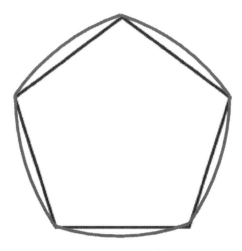

Figure 4.9:   Reuleaux Pentagon

Introducing probability

The mathematics field of probability offers parents an intriguing opportunity to entertain their children with things that are somewhat counterintuitive, yet instructive. Before getting into some of these examples it would be nice to explain to your child how this field evolved through a correspondence between two French mathematicians. (There will be more about these two famous mathematicians in the last chapter.)

The concept of probability plays a very important role in our society today. Yet it all began in 1654 when two famous French mathematicians, Blaise Pascal (1623–1662) and Pierre de Fermat (1601–1665) corresponded about the chances of winning in a then-popular dice game. The game consisted of tossing a pair of dice 24 times in determining whether to bet even money that at least one of those 24 tosses will result in a pair of sixes. It was believed then that getting a pair of sixes during these 24 throws was a good bet, yet the opposite resulted from the correspondence between these two famous mathematicians. They further corresponded about other such gaming outcomes, which eventually led to determining the mathematical field of probability, and which was formalized in 1657 by the first book published in this field by the Dutch mathematician Christian Huygens (1629–1695), entitled *De Ratiociniis in Ludo Aleae*. The continued study of probability has led to many unusual results, which we shall introduce in the following sections.

Friday the Thirteenth!

From a variety of sources, children tend to be exposed to the notion that the number 13 is usually associated with being an unlucky number. Buildings with more than thirteen stories, typically will omit the number 13 from the floor numbering. This is immediately noticeable in the elevator, where there is no button for 13. You can certainly think of other examples where the number 13 is associated with bad luck. This fear of the number 13 is often referred to as *triskaidekaphobia*. Among the more famous people in history who suffered from triskaidekaphobia are: Franklin D. Roosevelt, Herbert Hoover, and Napoleon Bonaparte.

You will likely recall that when the 13th of a month falls on a Friday, some folks believe that there would be bad luck on that day. This may derive from the belief that there were **thirteen** people present at "The Last Supper," which resulted in Jesus' crucifixion on a **Friday**. Do you think that the 13th comes up on a Friday with equal regularity as on the other days of the week? You may be astonished that, lo and behold, the 13th comes up more frequently on Friday than on any other day of the week.

This fact was first published by the American mathematician Bancroft H. Brown (1894–1974) in *American Mathematical Monthly* (1933, vol. 40, p. 607). He stated that the Gregorian calendar follows a pattern of leap years, repeating every 400 years. The number of days in one four-year cycle is $3 \cdot 365 + 366$. So, in 400 years there are $100(3 \cdot 365 + 366) - 3 = 146,097$ days. Note that the century year, unless divisible by 400, is not a leap year; hence the deduction of 3. This total number of days is exactly divisible by 7. Since there are 4800 months in this 400-year cycle, the 13th comes up 4800 times. Interestingly enough, the 13th comes up on a Friday more often than on any other day of the week. The following chart (Figure 4.10) summarizes the frequency of the 13th appearing on the various days of the week. Here we have a conundrum: the day that occurs most frequently is considered bad luck!

There are some people who were not consciously affected by the number **13**, but have had much of their lives involving this rather famous number.

Day of the week	Number of 13s	Percent
Sunday	687	14.313
Monday	685	14.271
Tuesday	685	14.271
Wednesday	687	14.313
Thursday	684	14.250
Friday	*688*	*14.333*
Saturday	684	14.250

Figure 4.10:   Frequency of the number 13 days of the week

For example, the famous German opera composer Richard Wagner was born in 18**13** (and the sum of the digits of this year is **13**), and his name consists of **13** letters. Wagner was first motivated towards his life work at a performance of Carl Maria von Weber's opera Der Freischütz on October **13**, 1822. He composed **13** operas. One of his operas, Tannhäuser, was completed on April **13**, 1845, and was first performed in Paris during his exile from Germany on March **13**, 1861. He spent **13** years in exile for political reasons. Wagner's last day in the city of Bayreuth, Germany, where he had built his famous opera house, was September **13**, 1882. His father-in-law the Hungarian composer Franz Liszt (1811–1886) saw Wagner for the last time in January **13**, 1883. Wagner died on February **13**, 1883, which just happens to be the **13**th year of the unification of Germany. Curiosities such as these bring another dimension of entertainment to numbers. You might search for other peculiarities regarding the number **13**, but they need not be negative as we see from the above illustration. This will also give kids an opportunity to search around and come back to you with some findings.

Unexpected Birthday Matches

We now of embark on an example that shows how probability can sometimes be very counterintuitive. This could actually be a rather pleasant experience for parent and child to undertake. As you can see as we go through this topic, you can have your child collect the birthdates of about 35 of his or her friends or classmates and then do what we are going to do here and then observe what results follow.

In our everyday lives, we might consider the likelihood or probability of something happening, or not happening. This often goes under the heading of how probable or how likely something is to happen. The topic of probability can explain many things. Yet, one of the most surprising results in mathematics — and one that is quite counterintuitive — is the question of how likely is it for two people to have the same birth date (just month and year). When you find out the likelihood of this happening, it will surely upset your sense of intuition. On the other hand, it is one of the best ways to convince the uninitiated about the "power" of probability.

Let us suppose that your child is in a group with 35 other people. What do you think the chances (or probability) are of two people in the group having the same birthdate (month and day, only)? Intuitively, one usually begins to think about the likelihood of 2 people having the same date out of a selection of 365 days (assuming no leap year). Perhaps 2 out of 365? That would be a probability of $\frac{2}{365} = .005479 \approx \frac{1}{2}\%$. A rather minuscule chance.

Rather than our group of 35 people, let us consider the "randomly" selected group of the first 35 presidents of the United States. You might have a child look up the birthdates and then find to his or her astonishment that there are two presidents with the same birthdate:

the 11th president, James K. Polk (November 2, 1795), and
the 29th president, Warren G. Harding (November 2, 1865).

You may be surprised to learn that for a group of 35, the probability that two members will have the same birthdate is greater than 8 out of 10, or $\frac{8}{10} = 80\%$.

An ambitious child might be motivated to ask his or her teacher to help conduct an experiment by selecting 10 classes or groups in the school building of about 35 members in each group and then check for birthdate matches each group. For groups of only 30 people, the probability that there will be a match of birthdates is greater than 7 out of 10, or expressed in another way: in 7 of these 10 groups there ought to be a match of birthdates. What causes this incredible and unanticipated result? Can this really be true? It seems to go against our intuition.

To relieve you of your curiosity and perhaps giving you the opportunity to share this with your child, we will consider the situation in detail. Let us consider a group of 35 people. What do you think is the probability that one selected person matches his or her own birthdate? Clearly *certainty*, which we express as a probability of 1. This can be written as $\frac{365}{365}$.

The probability that another person in the group does *not* match the first person is $\frac{365-1}{365} = \frac{364}{365}$.

The probability that a third person does *not* match the first and second person is $\frac{365-2}{365} = \frac{363}{365}$.

The probability of all 35 people *not* having the same birth date is the product of these probabilities: $p = \frac{365}{365} \cdot \frac{365-1}{365} \cdot \frac{365-2}{365} \cdot \ldots \cdot \frac{365-34}{365}$.

Since the probability (q) that two people in the group *have* the same birthdate and the probability (p) that two people in the group do *not have* the same birthdate is a certainty, the sum of those probabilities must be 1. Thus, $p + q = 1$.

In this case, $q = 1 - \frac{365}{365} \cdot \frac{365-1}{365} \cdot \frac{365-2}{365} \cdot \ldots \cdot \frac{365-34}{365} \approx .8143832388747152$. In other words, the probability that there will be a birthdate match in a randomly selected group of 35 people is somewhat greater than $\frac{8}{10}$. This is quite unexpected when one considers there were 365 dates from which to choose. The motivated reader may want to investigate the nature of the probability function. Figure 4.11 may further enlighten you.

Notice how quickly "almost-certainty" is reached. With about 60 students in a room the chart indicates that it is almost certain (.99) that two students will have the same birthdate.

Number of people in group	Probability of a birthdate match
10	.1169481777110776
15	.2529013197636863
20	.4114383835805799
25	.5686997039694639
30	.7063162427192686
35	.8143832388747152
40	.891231809817949
45	.9409758994657749
50	.9703735795779884
55	.9862622888164461
60	.994122660865348
65	.9976831073124921
70	.9991595759651571

Figure 4.11: Birthdate probabilities

Although it may be a somewhat sensitive topic, you might want to have your child take the list of the first 35 presidents, and see if there are any matches on death dates. Were you or your child to do this with the death dates of the first 35 presidents, you would notice that two died on March 8th (Millard Fillmore in 1874 and William H. Taft in 1930) and three presidents died on July 4th (John Adams and Thomas Jefferson in 1826, and James Monroe in 1831). How curious that the only two signers of the Declaration of Independence who went on to become presidents (Adams and Jefferson) died on the very same day — the date that the Declaration of Independence was ratified!

You might want to stress to your child that this astonishing demonstration should serve as an eye-opener about the inadvisability of relying too much on intuition, and becoming aware of how mathematics is truly a part of our lives.

Selecting Clothes

Here is a topic that you could experiment with your child to see if the mathematical logic holds up. However, it is important to stress the fact that this common-sense reasoning is an integral part of the mathematical skills that one should capture at an early age. We begin by introducing the task of selecting clothes, where we hope to show that it can be seen as a mathematical problem to be solved. One might also consider this as an exercise in logical thinking. Here we see an example that shows how reasoning with extremes is a particularly useful strategy to solve problems. It can also be seen as a "worst-case scenario" strategy. The best way to get to appreciate this kind of thinking is through an example. Therefore, we will consider the following problem.

> In a drawer, there are 8 blue socks, 6 green socks, and 12 black socks. What is the least number of socks that Charlie must take from the drawer in a dark room — where he cannot see the colors — to be certain that he has selected two socks of the same color?

The problem does not specify which color, so any of the three colors would satisfy his selection. To solve this problem, you might reason from the perspective of a "worst-case scenario." Suppose Charlie is very unlucky and on his first three picks, he picks one blue sock, one green sock, and then one black sock. He now has one of each color, but no matching pair. Although, he might have picked a matching pair on his first two selections, we need to determine how many socks, you must select to be certain of having a matching pair. As soon as he now picks the fourth sock, he must match one of his first three picks, and therefore, would have a pair of the same color.

Let's go back to the sock drawer and consider an alternative problem situation as expressed in the following problem:

> **In a drawer, there are 8 blue socks, 6 green socks, and 12 black socks. What is the least number of socks that Max must take from the drawer — without looking — to be certain that he has selected 2 black socks?**

Although this problem appears to be similar to the previous one, there is one important difference. In this problem, a specific color has been required. Here, Max is required to get a pair of black socks. Again, let's use deductive reasoning and construct the "worst-case scenario." Suppose Max first picks all of the eight blue socks. Next, he picks all six green socks. Still not one black sock has been chosen. Max now has selected 14 socks in all, but none of them is black. However, the next two socks he picks must be black, since there are only black socks remaining in the drawer. In order to be certain of picking two black socks, Max must select $8 + 6 + 2 = 16$ socks in all.

Here's a situation which many children face as they approach their teenage years, and yet, it provides us with a lovely example of mathematical reasoning. There are other considerations when selecting clothes to wear, which require what is known in mathematics as the *fundamental accounting principle*. Consider the following. Suppose we are traveling and have taken along 5 shirts, 3 pairs of pants, and 2 jackets. The question then is how many different outfits (or combinations) can we make from the clothes we have taken along? The reasoning is that for each of the 3 pairs of pants we can have any one of 5 shirts, which means there are 15 possible arrangements of these 2 items. Then for each of the 15 possible arrangements we can select one of the 2 jackets, providing us with 30 possible outfits. What the fundamental accounting principle tells us is that we merely have to multiply the 3 numbers to get the total number of outfits.

Playing Billiards Cleverly

Children who play the game of billiards, and play it well, have a natural talent for knowing how to hit a ball against the cushion so that it bounces off and lands at a predetermined spot. For those who have not yet played the game of billiards, this might be an opportunity for them to see how mathematics can be useful in sports. Through very simple mathematics, we can find the desired point on the cushion, which would allow the ball to ricochet to the aimed spot. But before we employ a mathematical strategy, which will justify the technique, let's discuss the physical aspects involved.

Imagine that you have the cue ball located somewhere on the table as shown in Figure 4.12. Your objective is to hit the cue ball against the side cushion and have it ricochet, so that it would hit the target ball, also shown in Figure 4.12. The technique is very simple. All we need to do is to place a mirror against the side cushion, and aim the cue ball at the mirror, shooting it to the point at which the target ball is seen in the

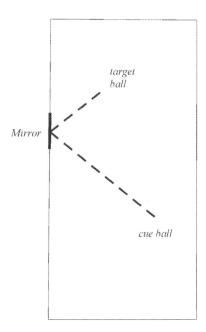

Figure 4.12: Billiard table

mirror. Of course, once you have marked that point, remove the mirror (so that you don't break it) and aim the ball to that point. Before progressing to an explanation make sure that your child truly understands what is going on.

The explanation that justifies this technique uses a concept from *geometric transformations*. In this case the transformation is a *reflection*. To reflect a given point in the given line, we simply locate the new point along the perpendicular to the given line from the given point and at an equal distance from the line (but on the other side of the line) as shown in Figure 4.13.

We are now ready to reflect the target ball in the side cushion, which is geometrically equivalent to seeing the reflection of the target ball in the mirror, which was placed against the side cushion. We then aim the ball at the mirror-reflection of the target ball with the straight-line segment shown in Figure 4.14. The ball will then ricochet off the cushion at the point *P* and hit the target ball as desired.

At this point you might explain to your child that we have sent the ball on the shortest path from the starting point to the cushion and to the

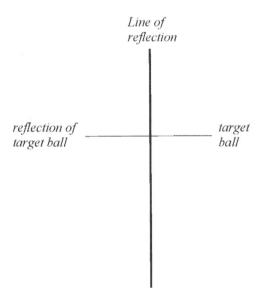

Figure 4.13:   Point reflected in the line

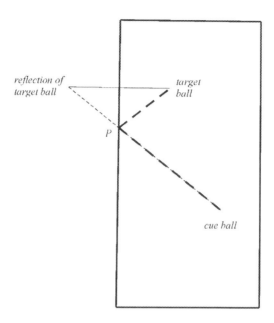

Figure 4.14: Billiard table showing point reflection

desired end position. We can easily show that this is the shortest path from the cue ball to the cushion to the target ball. This can be done comparing it to any other path. To do that we would choose any other point Q on the cushion, and compare the distance traveled from the cue ball's starting point to the point Q and then to the target ball with the distance of the ball's previous path through point P. (See Figure 4.15) Since triangle TPR is isosceles and $PR = PT$, the path from point C to P to T, or as stated geometrically $CP + PT$ is equal to $CP + PR$, Similarly, $CQ + QT = CQ + QR$. However, we know that the shortest distance between two points is a straight line, therefore, $CP + PR < CQ + QR$. Or stated verbally $CP + PR$ is shorter than any other path from the cue ball to the cushion to the target ball.

For parents who wish to take the billiard example one step further, namely, by hitting two cushions before hitting the target ball, we would need to reflect the target ball in the two cushions and then follow that path, as shown with dashed lines in Figure 4.16. That is, first reflect the target ball in one cushion, then reflect this reflection of the target ball in the

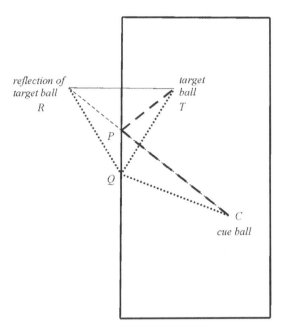

Figure 4.15:   Billiard table showing the shortest distance

extension of the other cushion, and then determine the points *A* and *B* noted in Figure 4.16.

The same effect can be obtained by reflecting each of the two balls in their respective cushions and then joining the two reflections to determine the points *A* and *B*, which are the points of contact on the cushions — as shown in Figure 4.17.

Although the geometric procedure we are using may not be practical to use during the billiard game, it does justify how using one or two mirrors can determine the points of reflection, which are the points on which the cue ball would need to hit in order to reach the desired target ball. This latter discussion might be a bit beyond where most students' interests would lie, however, parental judgment might find it a useful extension for enrichment.

Referring back to our initial discussion of reflections, we can also use this concept to solve a home construction problem. This might also be of interest to children as they usually try to look for practical applications. Supposing

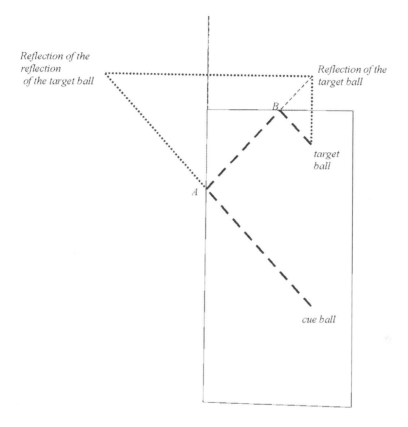

Reflection of the reflection of the target ball

Reflection of the target ball

target ball

A

B

cue ball

Figure 4.16: Billiard table showing double reflection

you would like to install an electric outlet along the wall, knowing that to this outlet two lamps will need to be connected. However, we would like to use the least amount of wire to connect these two lamps to the outlet. How might we locate this ideal point along the wall. Let's consider the situation shown in Figure 4.18.

One way of locating this ideal point for the electrical outlet is to find the reflection of lamp 1 in the wall, and then join that reflection point with a straight line to lamp 2. The point at which that line segment intersects the side of the wall is the point at which the outlet should be placed, since, as we saw earlier, that is the minimum distance point from lamp 1 to the electrical outlet to lamp 2.

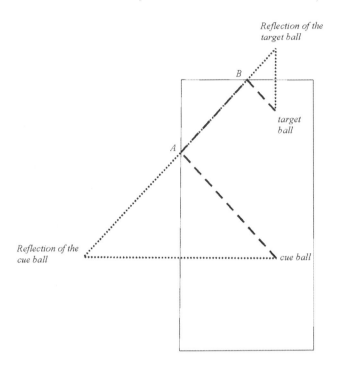

Figure 4.17:   Billiard table showing double reflection — an alternative

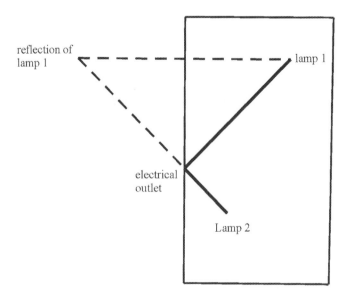

Figure 4.18:   Ideal wiring of two lamps in a room

Measuring the Earth

Parents' responsibility for their child's mathematics learning does not end at the elementary grades and should continue on through the middle grades and perhaps even on into high school. There are times where you can see topics which are typically not covered in the curriculum and that could be of general interest not only to children, but for adults as well. Let's consider some of these now, such as measurements of and on the Earth.

You might begin by asking your child if he or she has ever stood at the beach and wondered how far can you see to the horizon line, that is, the separation between sky and the water? Of course, this depends to some degree — although be it minimal — on the height of the observer. With simple geometry, we can calculate that distance. However, before we determine the distance to the horizon line, let's see how one could possibly measure the size of the Earth, which for our purposes here, we will consider to be a perfect sphere, even though we know that there is a slight difference in the diameter measured between the North Pole and South Pole (7,898 miles) and the equatorial diameter (7,926 miles), for an average diameter of 7,912 miles. Using the formula for the circumference of a circle πd, we get an average circumference of about 24,856 miles.

With the modern instrumentation, measuring the Earth today is not terribly difficult, but thousands of years ago this was no mean feat. Remember the word "geometry" is derived from "Earth measurement." Therefore, it is appropriate to consider this issue of Earth measurement in one of its earliest forms. One of these measurements of the circumference of the Earth was made by the Greek mathematician Eratosthenes (276 BCE–194 BCE) in about 230 BCE. His measurement was remarkably accurate, being less than 1 percent in error. To make this measurement, Eratosthenes used the relationship of alternate-interior angles of parallel lines, to which students are introduced in high school geometry or sometimes in middle school.

As the librarian of Alexandria, Eratosthenes had access to records of calendar events. He discovered that at noon on a certain day of the year in the town of Syene (now called Aswan, Egypt) on the Nile River, the sun was directly overhead. As a result, the bottom of a deep vertical well was entirely lit and a vertical pole, being parallel to the rays hitting it, cast practically no shadow.

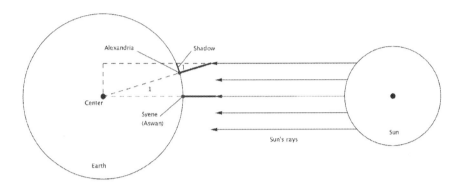

Figure 4.19:   Eratosthenes measuring the earth

At the same time, however, a vertical pole in the city of Alexandria, Egypt did cast a shadow. When that day arrived again, Eratosthenes measured the angle ($\angle 1$ in Figure 4.19) formed by such a pole and the ray of light from the sun going past the top of the pole to the far end of the shadow. He found it to be about $7°12'$, or $\frac{1}{50}$ of $360°$.

Have your child accept that we are assuming the rays of the sun to be parallel. Thus, Eratosthenes knew that the angle at the center of the Earth must be congruent to $\angle 1$, and hence, must also measure approximately $\frac{1}{50}$ of $360°$. Since Syene and Alexandria were almost on the same meridian or longitudinal line, Syene must be located on the radius of the circle, which was parallel to the rays of the sun. Eratosthenes therefore, deduced that the distance between Syene and Alexandria was $\frac{1}{50}$ of the circumference of the Earth. The distance from Syene to Alexandria was believed to be about 5,000 Greek *stadia*. A *Stadium* was a unit of measurement equal to the length of an Olympic or Egyptian stadium. Therefore, Eratosthenes concluded that the circumference of the Earth was about 250,000 Greek stadia, or about 24,660 miles. This is very close to modern calculations, which have the circumference at 24,856 miles. So how's that for some *real geometry*! You just had an opportunity to provide your child with a topic that is probably not going to be covered in school. However, it is an important topic and will give your child an opportunity to recognize how mathematics seems to crop up everywhere.

Let's consider our original problem, that of determining the distance of the horizon. Once again, we will work with the assumption that the Earth

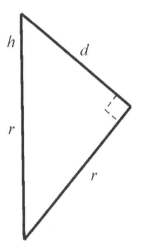

Figure 4.20

is a perfect sphere. We will be standing on a beach on a perfectly clear day, ignoring any light refraction that may occur while passing through the atmosphere, and our eyes will be at a height of h above the sea level and the distance to the horizon will be designated by d, with r representing the radius of the Earth. (See Figure 4.20).

Applying the Pythagorean theorem to the right triangle shown in Figure 4.20, we get the following: $(r+h)^2 = d^2+r^2$. Compared to the radius of the Earth, our height above sea level is negligible, so that we can safely ignore h^2, without too much loss of accuracy, when we square the binomial above, to get $(r + h)^2 = r^2 + h^2 + 2rh \approx r^2 + 2rh$. By then substituting this value for the binomial, we get $r^2 + 2rh = d^2 + r^2$, which then simplifies to: $d = \sqrt{2rh}$. Let's assume that a person's eyes are 6 feet above the ground, which would give us $h = 6$ feet or $\frac{6}{5,280}$ miles. Therefore, with the Earth's average radius of 3,956 miles, we get $d = \sqrt{(2)(3,956)\left(\frac{6}{5,280}\right)} = \sqrt{8.99} \approx 2.9$ miles. Naturally, if you are standing on a higher platform, such as a lifeguard's seat, the distance you can see to the horizon will be considerably longer. For example, when one looks at the horizon from the magnificent ledge at an elevation of 2250 feet, near the village of Haines Falls in the Catskill Mountains of New York State, where previously the famous Catskill Mountain House once stood (and where three United States presidents vacationed — Ulysses S Grant, Chester A. Arthur, and Theodore Roosevelt), according to our formula you should be able to see the horizon line at a distance of about 58 miles. To consider an extreme, suppose you are at the top of Mount Everest, at an elevation of 29,029 feet. From there, the distance to the horizon would be approximately 208 miles. You now have a method of determining whether the lyrics by Alan Jay Lerner were correct in his musical *On a Clear Day You Can See Forever*. Now we know that he was exaggerating a bit!

Navigating the Globe

We offer parents here a rather nice little conundrum that at first mentioning appears to be impossible to solve, but after some thought a clever solution evolves. Leading your child through this requires patience and slow presentation so that he or she can understand exactly what is required.

There are entertainments in mathematics that stretch (gently, of course) the mind in a very pleasant and satisfying way. As we now "navigate" the Earth, we will begin with a popular puzzle question that has some very interesting extensions, and which will help us as we traverse the globe. Here we will be required to do some "out of the box" thinking, with the hope that we will leave your child with some favorable lasting effects. Let's consider the following question proposed to your child: Where on Earth can you be so that you can walk: one-mile **south**, then one-mile **east**, and then one-mile **north** and end up at the starting point?

Perhaps after a few attempted starting points, your child may come up with the right answer, which is the North Pole, of course! To test this answer, try starting from the North Pole (shown in Figure 4.21) and traveling one mile south and then going one mile east, which takes you along a latitudinal line — one that remains equidistant from the North Pole, at a distance of one mile from the Pole. Then by traveling one mile north you get back to where you began, the North Pole.

At this point if your child truly understands the solution to this problem, he or she will feel a sense of completion. Yet, as an interested parent you might want to ask your child, if there are other such starting points, where we can take the same three directional "walks" and end up at the starting point? The answer, surprising enough, is *yes*. One set of starting

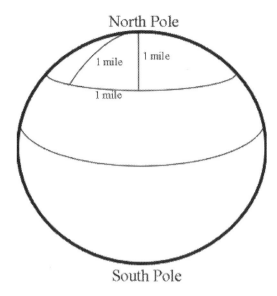

Figure 4.21: Sphere of the earth

points is found by locating the latitudinal circle, which has a circumference of one mile, and is nearest the South Pole. (See Figure 4.22) From this circle walk one mile north (along a great circle, which is a circle whose center is at the center of the sphere) and form another latitudinal circle. Any point along this second latitudinal circle will qualify. Let's try it.

Begin on this second latitudinal circle, which is one mile farther north. Walk one mile south, which takes you to the first latitudinal circle. Then walk one mile east, which takes you exactly once around the circle, and then one mile north, which takes you back to the starting point.

Now have your child consider the latitudinal circle closer to the South Pole, the one we would walk along, and which would have a circumference of one-half mile. We could still satisfy the given instructions, yet this time walking around the circle *twice*, and get back to our starting point on a latitudinal circle that is one mile farther north. If the latitudinal circle nearer to the South Pole had a circumference of one-quarter mile, then we would merely have to walk around this circle *four* times to get back to the starting point of this circle and then go north one mile to the original starting point.

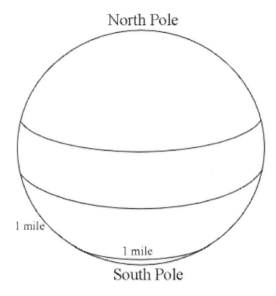

Figure 4.22:   Sphere of the earth

At this point your child might be ready to take a giant leap to a generalization that will lead to many more points that satisfy the original stipulations. Actually, an infinite number of points! This set of points can be located by beginning with the latitudinal circle, located nearest the South Pole, that has a $\frac{1}{n}th$ -mile circumference, so that a one-mile walk east will take you back to the point on the circle at which you began your walk on this latitudinal circle (since a one-mile walk on this circle means walking around this circle n times). The rest is the same as before, that is walking one mile south and then later one-mile north. Is this possible with latitude circle routes near the North Pole? Yes, of course! As you can see, mathematics can be quite helpful when navigating the globe, and once again you as a parent have the ability to motivate your child towards mathematics by using an example that is pleasingly logical and more than likely not presented in the school.

Coloring a Map

Here we give parents an opportunity to apply mathematics to a very common issue, yet one that leads to a new branch of mathematics, which is clearly not part of the school curriculum. Everyone is aware of what a map looks like, but rarely do we consider that when a map is designed, the coloring of the map's regions is an important consideration. Therefore, parents can ask that child if they ever wondered how a map is colored? Aside from deciding which colors should be used, the question might come up regarding the number of colors that are required for a specific map so that there will never be the same color on both sides of a border between countries. Well, mathematicians have determined the answer to this question, namely, that one will never need more than four colors to color any map, regardless of how many borders or contorted arrangements the map presents. For many years, the question as to how many colors are required was a constant challenge to mathematicians, especially those doing research in topology, which is a branch of mathematics related to geometry, where figures discussed may appear on plane surfaces or on three-dimensional surfaces. The topologist studies the properties of a figure that remain the same *after* the figure has been distorted or stretched according to a set of rules. A piece of string with its ends connected may take on the shape of a circle, or a square, which is all the same for the topologist. In going through this transformation, the order of the "points" along the string does not change. This retention of ordering has survived the distortion of shape, and it is this property that attracts the interest of topologists.

Throughout the 19th century it was believed that five colors were required to color even the most complicated looking map. However, there was always strong speculation that four colors would suffice. It was not until 1976 that the mathematicians Kenneth Appel (1932–2013) and Wolfgang Haken (1928–) "proved" that four colors were sufficient to color any map. However, unconventionally they used a high-powered computer to consider all possible map arrangements. It must be said, that there are still mathematicians who are dissatisfied with their "proof," since it was done by computer and not in the traditional way "by hand" with logical reasoning. Previously, it was considered one of the famous unsolved problems of mathematics, but it is now considered solved. Let us now inspect this problem by considering various maps, and the number of colors required to color them in such a way that no common boundary of two regions uses the same color on both of the sides. This is clearly a requirement for coloring any map.

Suppose we consider a geographic map that has a configuration analogous to that shown in Figure 4.23.

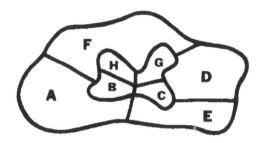

Figure 4.23: A sample map

 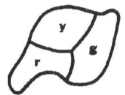 

Figure 4.24: Some sample maps

Here we notice that there are eight different regions indicated by the letters shown. Suppose we list all regions that have a common boundary with region H, and regions that share a common vertex with region H. The regions designated by the letters B, G, and F share a border with region H. The region designated by the letter C shares a vertex with the region designated by the letter H.

Remember, a map will be considered correctly colored when each region is completely colored and two regions that share a common boundary have different colors. Two regions sharing a common vertex may also share the same color. Let's consider coloring a few maps (Figure 4.24) to see various configurations that a map can have and that requires not more than three colors. (**b**/blue; **r**/red; **y**/yellow; **g**/green). You might have your child create a few other maps and experiment with a coloring arrangement.

The first map in Figure 4.24 can be colored in two colors: yellow and red. The second map required three colors: yellow, red and green. The third map has three separate regions, but only requires two colors, red and green, since the innermost territory does not share a common border with the outermost territory. It would seem reasonable to conclude that, if a three-region map

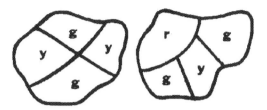

Figure 4.25:   More sample maps

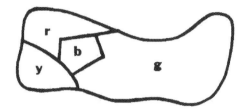

Figure 4.26:   A nap that requires four colors

can be colored with less than three colors, a four-region map can be colored with less than four colors. Let's consider such a map.

The left-side map shown in Figure 4.25 has four regions and requires only two colors for correct coloring. Whereas, the right-side map in Figure 4.25 also consists of four regions, but requires three colors for correct coloring.

We should now consider a map that requires four colors for proper coloring of the regions. Essentially, this will be a map, where each of the four regions shares a common border with the other three regions. One possible such mapping is shown in Figure 4.26.

If we now take the next logical step in this series of map-coloring challenges, we should come up with the idea of coloring maps involving five distinct regions. It will be possible to draw maps that have five regions, which require two, three, or four colors to be colored correctly. The task of drawing a five-region map that *requires* five colors for correct coloring will be impossible. This curiosity can be generalized through further investigation of other maps and should convince your child that the idea that any map, on a plane surface, with any number of regions, can be successfully colored with four or fewer colors. You might want to challenge your child to conceive of a map of any number of regions that requires more than four colors, where no two regions with a common border share the same color. They will soon see that they are unable to produce such a map. This is an excellent way for them to be convinced about the power of this result.

This challenge has remained alive for many years, challenging some of the most brilliant minds, but as we said earlier, the issue has been closed by the work of the two mathematicians, Appel and Haken. There are still many conjectures in mathematics that have escaped being proved. Here, at least, we have a conjecture that is closed. One such open conjecture, is the famous Goldbach conjecture, which states that every even integer greater than 2 can be expressed as the sum of two prime numbers. It might be fun for your child to begin the list of these even numbers along with the sum of the two prime numbers equal to each of these even numbers.

Crossing the Bridges

Parents often take their children for a walk. There are times when one tries to figure out the distance of a walk, or the number of lampposts one passes, or the number of bridges one crosses along the path. There is a famous problem in mathematics that stems from an age-old conundrum that fascinated folks in Europe for many years. Yet, this story cannot only fascinate children, but also gently introduce them to a relatively new field of mathematics involving the study of networks. We begin with a little bit of a historic background so that your child will become fascinated by the problem that faced generations of Europeans.

In the eighteenth century and earlier, when walking was the dominant form of local transportation, people would often count particular kinds of objects they passed. One such was bridges. Through the eighteenth century the small Prussian city of Königsberg (today called Kaliningrad, Russia), located where the Pregel River forms two branches, was faced with a recreational dilemma: Could a person walk over each of the seven bridges *exactly once* in a continuous walk through the city? The residents of the city had this as a recreational challenge, particularly on Sunday afternoons. Since there were no successful attempts, the challenge continued for many years.

This problem provides a wonderful window into networks, which is also referred to as graph theory, an extended field of geometry, which gives us a renewed view of the subject. To begin we should present the problem. In Figure 4.27 we can see the map of the city with the seven bridges highlighted.

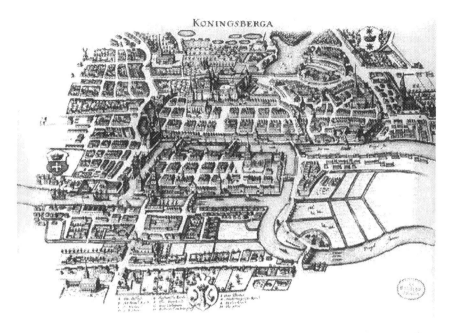

Figure 4.27: A map of Königsberg

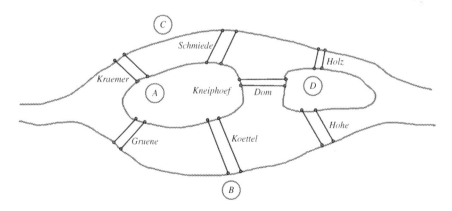

Figure 4.28: A simplified sketch of the Königsberg bridges

In Figure 4.28, we will indicate the island by A, the left bank of the river by B, the right bank by C, and the area between the two arms of the upper course by D. If we start at Holz and walk to Schmiede and then through Honig, through Hohe, through Köttel, through Grüne we *will* never cross Krämer. On the other hand, if we start at Krämer and walk to Honig, through

Hohe, through Köttel, through Schmiede, through Holz we will never travel through Grüne.

In 1735, the famous Swiss mathematician Leonhard Euler (1707–1783) proved mathematically that this walk could not be performed. The famous Königsberg Bridges Problem, as it became known, is a lovely application of a topological problem with networks. It is very nice to observe how mathematics — used properly — can put a practical problem to rest.

Before we embark on the problem we ought to become familiar with the basic concepts involved. Toward that end, have your child try to trace with a pencil each of the following configurations without missing any part and without going over any part twice. Make sure your child keeps count of the number of arcs or line segments, which have an endpoint at each of the points: *A, B, C, D, E*.

Configurations, called networks, such as the five figures shown in Figure 4.29 are made up of line segments and/or continuous arcs. The number of arcs or line segments that have an endpoint at a particular vertex, is called the *degree* of the vertex.

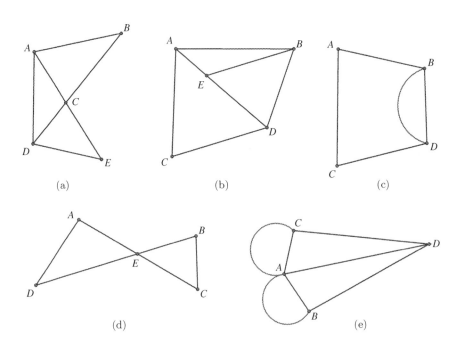

Figure 4.29:   Various networks

It would probably be fun to have your child try to trace these networks without taking the pencil off the paper and without going over any line more than once. He or she should notice two direct outcomes. The networks can be traced (or traversed) if they have

(1) all even degree vertices, or
(2) exactly two odd degree vertices.

The following two statements summarize this finding:

1. There is an even number of odd degree vertices in a connected network.
2. A connected network can be traversed, only if it has at most two odd degree vertices.

Network Figure 4.29a has five vertices. Vertices B, C, E are of even degree and vertices A and D are of odd degree. Since Figure 4.29a has exactly two odd degree vertices as well as three even degree vertices it is traversable. If we start at A then go down to D, across to E, back up to A, across to B, and down to D we have chosen a desired route.

Network Figure 4.29b has five vertices. Vertex C is the only even degree vertex. Vertices A, B, E, and D, are all of odd degree. Consequently, since the network has more than two odd vertices, it is not traversable.

Network Figure 4.29c is traversable because it has two even vertices and exactly two odd degree vertices.

Network Figure 4.29d has five even degree vertices and, therefore, can be traversed.

Network Figure 4.29e has four odd degree vertices and *cannot* be traversed.

The Königsberg Bridge Problem is the same problem as the one posed in Figure 4.29e. Let's take a look at Figure 4.29e and Figure 4.28 and note the similarity. There are seven bridges in Figure 4.28 and there are seven lines in Figure 4.29e. In Figure 29e each vertex is of odd degree. In Figure 4.28 if we start at D we have three choices, we could go to Hohe, Honig, or Holz. If in Figure 4.29e we start at D we have three line paths to choose from. In both figures, if we are at C we have either three bridges we could go on (or three lines). A similar situation exists for locations A and B in Figure 4.28 and vertices A and B in Figure 4.29e. We can see that this network cannot be traversed.

By reducing the bridges and islands to a network problem we can easily solve it. This is a clever tactic to solve problems in mathematics. You ask your child to try to find a group of local bridges in your region to create a similar challenge, and see if the walk is traversable. This problem and its network application is an excellent introduction into the field of topology. Getting your child actively involved with the various experiments, which we described above would ensure a genuine understanding on his or her part and perhaps the one step towards motivating your child appreciating an aspect of mathematics that would otherwise not be exposed to him or her.

Should your child have been interested in the above problem, we offer here an analogous situation that takes place in the house. Here we can apply the technique we just developed for determining the traversability of a network to the famous *five-bedroom-house* problem. Let's consider the floor plan of a house with five rooms as shown in Figure 4.30.

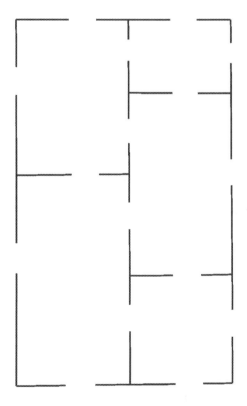

Figure 4.30:   Floorplan of the house with five rooms showing doorways

Each room has a doorway to each adjacent room and a doorway leading outside the house. The problem is to have a person start either inside or outside the house and walk through each doorway exactly once. You will realize that, although the number of attempts is finite, there are far too many ways to make a trial-and-error solution practical.

Figure 4.31 shows various possible paths joining the five rooms *A, B, C, D,* and *E* and the outside area *F*.

As before, our question can be answered by merely determining if this network is traversable. In Figure 4.31, we notice that we have marked the vertices with the letters *A, B, C, D, E* and *F*. We notice that 4 vertices are of odd degree, and 2 vertices are of even degree. Since there are not exactly 2 or 0 vertices of odd order, this network cannot be traversed. Therefore,

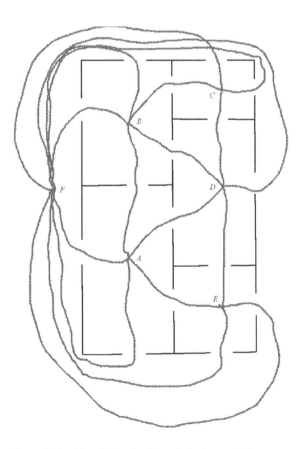

Figure 4.31: Traveled paths through the house of five rooms

the five-room-house problem does not have a solution path, which would allow walking through each doorway exactly once. As you can see, even in choosing paths for travel, mathematics seems to provide a solution to our question. Essentially, the message that you would want to bring to your child is there is lots of mathematics beyond what is presented in school, although what you learn in school is necessary to appreciate an understanding of the applications well beyond the curriculum.

Chapter 5

Fun with Mathematics

Some people find it strange to say that mathematics can be fun. Actually, the majority of folks in our culture is oftentime proud to have been weak in mathematics and still are enjoying the good life. Unfortunately, it is sometimes considered a badge of honor to say: "I was lousy at mathematics at school and didn't find it fun, but nonetheless I am a happy person." This is something that needs to be changed right from birth, so that a child encounters mathematics from a positive vantage point, namely, by saying, "Math really is fun!"

In this chapter, we will try to persuade parents that not only will they learn to enjoy the peculiarities, the novelties, the amazements, and the curiosities embedded in the field of mathematics, but we hope to convince them that mathematics can be fun. This belief is something that needs to be transmitted to children of all ages. Naturally, one needs to determine what is appropriate at the various ages. This is something that we can hint at, but the ultimate decision rests with the parents. There are times when it is wise to challenge students at younger ages, as well as showing older students simple things that they can embrace and share with their friends — and even their teachers. First, we need to provide parents with ideas that they themselves find interesting and fun. Then it is most important for parents to sort out what they believe suits their child best and then present it in a calm and convincing fashion — showing their own genuine interest in the topic — and thereby motivating their child to join those who find the topic exciting and fascinating.

Let us now take a small journey through a plethora of short, but entertaining, topics that can be used at various grade levels. We leave it to parents to decide what is appropriate for their children. Sometimes it's worth challenging students, other times it's just nice to show them the beauty of mathematics — as well as its power. Naturally, there are specific factors that parents need to take into consideration when deciding which of the "fun" items presented here will be most appropriate for their child. Among other things to consider is the environment in which the child flourishes, the child's readiness to grasp the topic or concept, so that it not only is meaningful but also entertaining. And above all, the way in which parents present topics or ideas to their children is critical. As we said previously, these ideas need to be presented in a friendly entertaining fashion, where the parents exhibit their own enthusiasm for the idea presented. It is sometimes advisable not to just tell the child about this "fun item," but rather allow them to discover it and take genuine ownership of it.

Entertaining number curiosities

Let's begin with a rather simple, yet entertaining, peculiarity in arithmetic. There are numbers that have unusual properties, such as all two-digit numbers ending in 9. They are equal to the product of the digits *plus* the sum of their digits. We can test that with a few such two-digit numbers:

For the number **39,** if we take the product and the sum of its digits we get the following:

$$(3 \times 9) + (3 + 9) = 27 + 12 = \mathbf{39},$$

and for the number **79,** we get

$$(7 \times 9) + (7 + 9) = 63 + 16 = \mathbf{79}.$$

You might want to have your child check out other two-digit numbers with the units digit 9.

Here is another number peculiarity that you might want to show your child — one that can provide some enjoyable practice for multiplication. The number 48 has a distinct property that the product of its proper factors is equal to 48^4. The list of its proper factors (which excludes 1 and 48) is: 2, 3, 4, 6, 8, 12, 16, 24. The product of all these factors is:

$$2 \times 3 \times 4 \times 6 \times 8 \times 12 \times 16 \times 24 = 5,308,416 = 48^4.$$

It might be too frustrating to ask your child to find other such numbers, but verifying this relationship could be fun.

You might want to have your child use a calculator (or do it with paper and pencil) to find the value of 69^2 and 69^3. They should get the following:

$$69^2 = \textbf{4,761},$$
$$69^3 = \textbf{328,509}.$$

At first one might say "so what is so curious about these calculations?" Ask your child to look at these two results (bold). And see if they can discover any thing unusual about these two sums. After a short while they should notice that the two results of the calculations represent all of the digits from 0 to 9. You might want to tell them that 69 is the *only number*, which when squared and cubed yields numbers using all the digits exactly once.

Here is a cute little trick you can show your child, which will not only impress him or her, but also provide for further thought. To begin, have your child to write the sequence of natural numbers in groups as shown below:

1
2, 3
4, 5, 6
7, 8, 9, 10
11, 12, 13, 14, 15
16, 17, 18, 19, 20, 21
22, 23, 24, 25, 26, 27, 28

Then have them cancel out every second group (as shown below), so that they are left with:

1
~~2, 3~~
4, 5, 6
~~7, 8, 9, 10~~
11, 12, 13, 14, 15
~~16, 17, 18, 19, 20, 21~~
22, 23, 24, 25, 26, 27, 28

If we take the sum of the first two remaining groups, we get:

$$(1) + (4 + 5 + 6) = 16 = 4^2.$$

If we take the sum of the first three remaining groups, we get:

$$(1) + (4 + 5 + 6) + (11 + 12 + 13 + 14 + 15) = 81 = 9^2.$$

If we take the sum of the first four remaining groups, we get:

$$(1) + (4 + 5 + 6) + (11 + 12 + 13 + 14 + 15) + (22 + 23 + 24 + 25 + 26 + 27 + 28) = 256 = 16^2.$$

Your child should be led to notice that the relationship of the three results that we obtained formed a pattern, where we actually arrived at 2^4, 3^4, and 4^4 and were we to continue we would have as our next result $25^2 = 5^4$. You might want to have your child continue this procedure to see if the pattern holds further on. It does!

Numbers consisting of repeating digits oftentime provide some interesting patterns. This can perhaps provide a child to seek other such patterns. As an example, have your child square the number 88. They will find that not only is 88 a repeating-digit number, but its square 7744 also consists of some repetition of digits. A parent's dramatization of this will help to motivate the child to seek other such numbers. One might also mention that the number 8 in China is considered a lucky number, which is why the Olympic Games held in China in 1988 began on August 8, 1988 at 8:08 PM, which can be written as 8/8/88 — 8:08.

There are times when the very simplest pattern or relationship can fascinate a child. Of course, it is up to the parent to select that which is appropriate considering interest and age. For example, consider this nice symmetry which you can use to entice your child:

$$10 = 1 + 2 + 3 + 4$$
$$100 = 1^3 + 2^3 + 3^3 + 4^3$$

Raising these numbers to the next odd power — the 5th power — still gives a nice result, namely, 1,300. Some curious surprises lie ahead when we take the sum of these first four natural numbers to the 7th power (the sum is 18,700).

You might like to challenge your child with a fun project. They should try to get the sum of the initial natural numbers — in order — to total 100. Trying this requires a bit of creative thinking, but could be a fun activity for kids at various ages. Here are two possibilities:

$$1 + 2 + 3 + 4 + 5 + 6 + 7 + (8 \times 9) = 100.$$
$$123 - 45 - 67 + 89 = 100,$$

or in using these numbers in the reverse order, we can get: $98 - 76 + 54 + 3 + 21 = 100$.

These activities at an early age — presented properly with appropriate motivation — can generate curiosity and encourages some logical thinking and, above all, make mathematics fun!

Some children like to get facts. So here are some for parents to present — in an interesting fashion to their child. You can tell your child that 81 is the only number whose square root is equal to the sum of its digits. That is, $\sqrt{81} = 9 = 8+1$. This is just a cute little curiosity. By the way, number 81 is also the smallest square, where the sum of its divisors $(1+3+9+27+81 = 121 = 11^2)$ is also a square.

You can also have fun with prime numbers (those that only have themselves and 1 as factors). For example, 113 is a prime number, since its only factors are 113 and 1. However, this particular prime is the smallest prime number where all the arrangements of the digits is also a prime number — that is, 131 and 311. Other such primes are 337 and 199.

Here is another interesting number property. We can always express a number as the sum of three other numbers. However, with the number 118, we can also express it as the sum of four arrangements of three numbers, and the amazing thing is that the product of each of these sets of three numbers is the same for all four groups of three, namely, 37,800. Take a look here:

$$15 + 40 + 63 = 118, \text{ and } 15 \times 40 \times 63 = 37,800,$$
$$14 + 50 + 54 = 118, \text{ and } 14 \times 50 \times 54 = 37,800$$
$$21 + 25 + 72 = 118, \text{ and } 21 \times 25 \times 72 = 37,800$$

$$18 + 30 + 70 + 118, \text{ and } 18 \times 30 \times 70 = 37,800$$

More amazing: this number 118 is the smallest number for which this can be done. You might want to challenge you child to come up with some other such arrangements for other numbers.

Sometimes, peculiarities so simple can be interesting. Take for example the fact that there are only two numbers, 4 and 11, where their squares increased by 4 will yield a cube.

$2^2 = 4$, then by adding 4, we get $4 + 4 = 8 = 2^3$

$11^2 = 121$, then by adding 4, we get $121 + 4 = 125 = 5^3$.

Here is another nice number relationship with which you can entertain your child. Look at the symmetry:

$13^3 - 3^7 = 2197 - 2187 = 13 - 3$

$5^3 - 2^7 = 125 - 128 = -(5 - 2)$

Let them appreciate the symmetry, where a pair powers that can achieve this pattern. Please do not ask your child to find another such pair of numbers, because no other such pair has yet been found!

Another cute number property is one, where the sum of the divisors is equal to a perfect square. You can either show your child some of these and ask him or her to find others, or simply have him or her prove if that is true for the following numbers: 3, 22, 66, 70, 81, and others. Let's use for an example the number 66. The sum of its divisors is:

$$1 + 2 + 3 + 6 + 11 + 22 + 33 + 66 = 144 = 12^2.$$

There are many cute number relationships with which you can entertain your child. We will continue to present a few of them here:

$135 = 1^1 + 3^2 + 5^3$, also

$175 = 1^1 + 7^2 + 5^3$, and

$518 = 5^1 + 1^2 + 8^3$, as well as

$598 = 5^1 + 9^2 + 8^3$.

You might ask your child to see if there are other such relationships, or even some such as:

$244 = 1^3 + 3^3 + 6^3$, and

$136 = 2^3 + 4^3 + 4^3$.

In the event that you might want to keep this to one number, you can do this with the following numbers:

$153 = 1^3 + 5^3 + 3^3$,

$370 = 3^3 + 7^3 + 0^3$,

$371 = 3^3 + 7^3 + 1^3$,

$407 = 4^3 + 0^3 + 7^3$.

We can take this even a step further by considering the four-digit numbers such as the following:

$1634 = 1^4 + 6^4 + 3^4 + 4^4$,

$8208 = 8^4 + 2^4 + 0^4 + 8^4$,

$9474 = 9^4 + 4^4 + 7^4 + 4^4$.

Please don't challenge your child to find other such four-digit numbers, since to date no others that have this property have been found!

As we had been showing above, there are times when children get motivated when they see unusual numerical relationships, and have an urge to see if they can find others on their own — oftentime to impress the parents and even show some of these relationships to their classmates as though they discover them themselves. Either way, it is another path to showing that mathematics can be fun.

As an extra activity, if you feel that your child knows what the factorial function is, then show him or her that $145 = 1! + 4! + 5!$. If you really want to blow their minds, show him or her that the next number for which this works is $40,585 = 4! + 0! + 5! + 8! + 5!$. (As a review, a factorial is defined as the product of all the integers up to including the number indicated, for example, $3! = 1 \times 2 \times 3 = 6$.) There are nicer relationships that could serve to motivate students further along their search for patterns, for example, $3^3 + 4^3 + 5^3 = 6^3$. Or you might want to present another nice pattern with which to impress your child: $7^0 = 7^1 + 7^2 + 7^3 = 400 = 20^2$. Now working with the number 20, if we take the product of all proper divisors of 20, we get 20^2, that is, $2 \times 5 \times 4 \times 10 = 400 = 20^2$, which is quite unusual!

When we stretch our realm of relationships we get a nice one to show your child as further evidence that math can be fun. There are apparently only two numbers that are a multiple of their reversal. That is, the number $1089 \times 9 = 9801$, and the number $2178 \times 4 = 8712$. It probably would not be wise to ask your child to find other numbers where the reversal is a multiple of the original number. So far none has been found! However, this might be a good opportunity to discuss with your child what it means that others have not yet been found. Namely, that brilliant mathematicians have searched for other such numbers using some very sophisticated methods.

For those who like some symmetry, take a look at numbers resulting from 13^2 and 14^2:

$13^2 = 169$ and $14^2 = 196$. Here is another nice pattern of reversal: $499 = 197 + 2$, and $197 \times 2 = 994$.

Continuing on with reversal numbers, consider pairs of two-digit numbers that also give us an entertaining pattern, such as $36 \times 84 = 3024 = 63 \times 48$, or perhaps $12 \times 42 = 504 = 21 \times 24$. You might challenge your child to find other such two-digit number pairs, of which there are 11 more be found.

While on the topic of digits have your child look at these two equations to see if he or she can find something unusual in each of these calculations.

$567^2 = 321,489$, and

$854^2 = 729,316.$

Hint: 567 and 854 are the only two numbers when squared yield a number that together with the original number represents all of our digits from 1 to 9 exactly once.

Having now been exposed to some of the most amazing number patterns and relationships, we can now embark on some shortcuts in calculating arithmetic situations. We can begin by considering the powers that the 2 factors of the base 10 possess. We shall assume that most children recognize that when a number ends in an even number, such as 2, 4, 6, 8, and 0, that the number is divisible by 2. Now, as parents, you can show your child that we can check to see if a given number is divisible by 4 by simply looking at the last 2 digits of the number and inspecting the number formed by the last 2 digits. If it is divisible by 4, then the original number itself is divisible by 4. We can extend this even further to consider divisibility by 8. This can be achieved by looking at the last 3 digits of any number considered as a separate number. If that number is divisible by 8, then the original number is divisible by 8. Depending upon the reaction from your child, you might want to take this one step further by showing that there is an analogous technique that can be used to check for divisibility by 5, 25, and 125. That is, if a number ends in a digit that is divisible by 5, such as 5 or 0, then we know that the original number is divisible by 5. Analogously, if the last 2 digits of the number forms a number that is divisible by 25, then the original number is divisible by 25. And similarly, if the last 3 digits of a number forms a number, which is divisible by 125, then the original number is divisible by 125.

Let's now continue our journey with some arithmetic tricks, or shall we say shortcuts. But before we do that, we should pay homage to two numbers that have a special place in arithmetic. They are the numbers on either side of the number 10, which is the base of our number system. They are the numbers 9 and 11. First of all, the reciprocals of these two numbers show a lovely pattern: $\frac{1}{9} = 0.111111\ldots$, and $\frac{1}{11} = 0.09090909\ldots$. Then to make this even more attractive show your child that $\frac{1}{99} = 0.01010101\ldots$. This is a nice way to introduce these important numbers — 9 and 11.

Let's continue with the number 9, and look at some of its oddities. For example, it is the only square number that can be expressed as the sum of two cubes: $9 = 1^3 + 2^3$. It can also be expressed as a sum of three consecutive factorials: $9 = 1! + 2! + 3!$, where, as we indicated earlier, factorial is defined as the product of all the integers up to including the number indicated, for example, $3! = 1 \cdot 2 \cdot 3 = 6$. There are also a number of arrangements involving the number 9. Such as $9^2 = 81$, and $8 + 1 = 9$.

We are now ready to look at some curiosities about the interesting number 9. You can impress your child by telling him or her that you can look at a number, and without doing division, determine if the number is divisible by 9. The trick for doing this is to determine if the sum of the digits of a given number is divisible by 9. If it is, then the number itself is divisible by 9. Let's take, for example, the number 48,645, and see if it is divisible by 9 — without actually doing the division. The sum of the digits of this number is $4 + 8 + 6 + 4 + 5 = 27$, which is divisible by 9; therefore, the original number, 48,645, is also divisible by 9. We can nicely extend this divisibility rule for the number 3, so that when a given number has the sum of its digits divisible by 3 then we can conclude that the number is divisible by 3. If we wish to determine if the number 12,345 is divisible by 3, we simply take the sum of its digits $1 + 2 + 3 + 4 + 5 = 15$, which is divisible by 3; therefore, we can conclude that the number 12,345 is also divisible by 3. If your child is wondering about the usefulness of this divisibility rule, where they would otherwise simply use a calculator, you might be able to give an example of having to split a certain sum of money amongst 3 friends to see if it can be evenly split without actually going through the division.

The number 9 also allows us to check arithmetic, such as addition and multiplication with a technique usually referred to as "casting out nines." This was exposed to the Western world by Leonardo of Pisa, better known today as Fibonacci. In his book *Liber Abaci* written in the year 1202, Fibonacci first introduced the Hindu-Arabic numerals, which we use today. There he also showed the Western world his fascination with the arithmetic calculations he experienced in the Islamic world. He introduced the system of "casting out nines" — which refers to a technique to determine if an arithmetic calculation result is possibly correct. Before introducing this technique to your child, you might want to actually apply it to an addition or multiplication task you may have assigned to him or her earlier.

The process requires subtracting a specific number of groups of 9 from this sum (or you might say, taking bundles of 9 away from the sum). Although this technique might come in handy, the nice thing about it is that it demonstrates a hidden magic in ordinary arithmetic. Yes, your child can check his arithmetic using this procedure. Before we discuss this arithmetic-checking procedure, we will consider how the remainder of a division by 9 compares to removing groups of 9 from the digit sum of the number. Let us find the remainder, when 8,768 is divided by 9. The quotient is 974 with a remainder of 2. This remainder can also be obtained by "casting out nines" from the digit sum of the number 8,768: This means that we will find the sum of the digits, and if the sum is more than a single digit we shall repeat the procedure with the obtained sum. In the case of our given number, 8,768, the digit sum is $8 + 7 + 6 + 8 = 29$). Since this result is not a single-digit number, we will repeat the process with the number 29. Again, the casting out nines procedure is used to get: $2 + 9 = 11$, and again repeating this procedure for 11, we get: $1 + 1 = 2$, which was the same remainder as when we earlier divided 8,768 by 9.

We believe that your child will be more impressed when we can now take this process of casting out nines to another application, that of checking multiplication. Perhaps it is best to see it applied, as with checking the following multiplication to determine if it is correct: $734 \times 879 = 645,186$. We can check this by division, but that would be somewhat lengthy. However, we can also see if this product could be correct by "casting out nines." To do that we will take each of the factors and the product and then add the digits of each number — continuing this process as before until a single digit results:

For 734: $7 + 3 + 4 = 14$; then $1 + 4 = 5$

For 879: $8 + 7 + 9 = 24$; then $2 + 4 = 6$

For 645,186: $6 + 4 + 5 + 1 + 8 + 6 = 30$; then $3 + 0 = 3$

Since the product $5 \times 6 = 30$, which yields 3 (by casting out nines: $3 + 0 = 3$), is the same as for the product (30), the answer could be correct.

For practice, we will do another casting-out-nines "check" for the following multiplication:

$56,589 \times 983,678 = 55,665,354,342$

For 56,589: $5 + 6 + 5 + 8 + 9 = 33$; $3 + 3 = 6$

For 983,678: $9 + 8 + 3 + 6 + 7 + 8 = 41$; $4 + 1 = 5$

For 55,665,354,342: $5 + 5 + 6 + 6 + 5 + 3 + 5$
$+ 4 + 3 + 4 + 2 = 48$; $4 + 8 = 12$; $1 + 2 = 3$

To check for possibly having the correct product: $6 \times 5 = 30$ or $3 + 0 = 3$, which matches the 3 resulting from the product digit.

A similar procedure can be used to check for the likelihood of a correct sum or quotient, simply by taking the sum (or quotient) and casting out nines, taking the sum (or quotient) of these "remainders" and comparing it with the remainder of the sum (or quotient). They should be equal, if the answer is to be correct.

The number 11, which is one greater than the number of our base 10, also has some nifty characteristics. From an aesthetic point of view, it is the only palindromic (reads the same in both directions) prime number with an even number of digits. However, kids get a kick out of being able to look at a number and determining whether it is divisible by 11 without actually doing the division; just by using the following technique. All one needs to do is to take the difference of the sums of the alternating digits and inspect the result. If the result is divisible by 11, then the original number was divisible by 11. Parents would be wise to do a few examples to make sure that their child understands this rather complicated sentence. Let's take one example here: Without doing the division, we wish to know if the number 46,915 is divisible by 11. The sum of the odd-positioned digits is: $4+9+5 = 18$, and the sum of the even-positioned digits is: $6 + 1 = 7$. Since the difference, $18 - 7 = 11$, which is surely divisible by 11, the original number is also divisible by 11. Once again, we suggest having your child do a few of these for practice.

There is also a rather cute way to multiply by 11 — mentally! This typically enchants most children at a variety of ages. Applying this technique to a two-digit numbers is rather simple. You merely add the digits and place this sum between the other two digits. If the sum turns out to be a two-digit number, then place the units digit of this sum between the two digits of the original number and carry the tens digit to the tens digit of the original number. To get comfortable with this useful technique, here are a few examples that you might want to show your child:

$$34 \times 11 = 3\ (3+4)\ 4 \text{ or } 3\ (7)\ 4 = 374$$
$$67 \times 11 = 6\ (6+7)\ 7 = 6\ (13)\ 7 = (6+1)\ (3)\ 7 = (7)\ (3)\ 7 = 737.$$

The more practice your child can have with this technique, the more easily he or she will be able to use it in the future. Just in case your child wants to know if this procedure can be extended beyond two-digit numbers, the answer is "yes." It might be best to show the following multiplication using this technique: Let's use the number 12,345 and multiply it by 11 with this procedure. Here we begin at the right-side digit and add every pair of digits going to the left: $1\ [1+2]\ [2+3]\ [3+4]\ [4+5]\ 5 = 135{,}795$. As was the case with the two-digit-number example, if the sum of two digits is greater than 9, then we place the units digit appropriately and carry/add the tens digit to the next digit to the left. Once again, practice will ensure that your child will incorporate this in his arsenal of arithmetic tools.

A technique for testing divisibility for the numbers 7, 11 and 13 at the same time is not something that is particularly useful, yet, it could be fun to demonstrate to children and then to have them check their results with a calculator. The procedure may sound a little complicated, but we will clear it up with an example. We begin by partitioning a large number in groups of three digits, beginning at the right side. We will then add first, third and fifth groups of digits, and subtract from this sum the sum of the second, fourth and sixth groups of digits. To illustrate this procedure let's determine if the number 76,834,758 is divisible by each of the three numbers 7, 11 and 13.

Adding the first and third group, we get $76 + 758 = 834$. From this sum (834) we now subtract the second group, to get $834 - 834 = 0$, which is divisible by each of the numbers 7, 11 and 13. As we said earlier, although this is not very useful, it could open a child's interest to investigate why this happens. A key to the answer could be to consider the product of the three numbers 7, 11, and 13, which is 1001.

There are times when you just want to have some fun with numbers and entertain your child without there being any deeper meaning behind. Although this is merely entertaining, it will be a breath of fresh air compared to the regular grind of the school curriculum. Some might say that there is a quirk in our decimal number system that manifests itself with the following loop. There isn't much you can do with it, other than to marvel at the amazing outcome — that should be enough! By the way, this is a nifty way to have your child practice subtraction — without using a calculator. So, here is the process of this miraculous loop.

Follow the steps below.

1. **Begin by selecting a four-digit number (except one that has all digits the same).**
2. **Rearrange the digits of the number so that they form the largest number possible. (That means write the number with the digits in descending order.)**
3. **Then rearrange the digits of the number so that they form the smallest number possible. (That means write the number with the digits in ascending order. Zeros can take the first few places.)**

4. Subtract these two numbers (obviously, the smaller from the larger).
5. Take this difference and continue the process, over and over and over, until you notice something "unusual" happening. Don't give up before something unusual happens.

Depending on the number you selected to start with, you will eventually arrive at the number **6,174** — perhaps after one subtraction, or after several subtractions. When you do, you will find yourself in an endless loop. Remember, that you began with an arbitrarily selected number. Although some readers might be motivated to investigate this further, others will just sit back in awe.

Here is an example of how this works with our arbitrarily selected starting number 3,203.

We will select the number 3,203.
The *largest* number formed with these digits is:　　3320
The *smallest* number formed with these digits is:　　0233
The difference is:　　3087

Now using this number, 3087, we continue the process:
The largest number formed with these digits is:　　8730
The smallest number formed with these digits is:　　0378
The difference is:　　8352

Again, we repeat the process.
The largest number formed with these digits is:　　8532
The smallest number formed with these digits is:　　2358
The difference is:　　6174

The largest number formed with these digits is:　　7641
The smallest number formed with these digits is:　　1467
The difference is:　　6174

This nifty loop was first discovered by the Indian mathematician, Dattathreya Ramachandra Kaprekar (1905–1986), in 1946.[1] We often refer

---

[1] Kaprekar announced it at the Madras Mathematical Conference in 1949. He published the result in the paper *Problems involving reversal of digits* in *Scripta Mathematica* in 1953; see also Kaprekar, D. R.: An Interesting Property of the Number 6174. Scripta Math. 15(1955), 244–245.

to the number 6174 as the *Kaprekar constant*. And so the loop is formed, since once you arrive at 6,174, you keep on getting back to 6,174. Remember, all this began with an arbitrarily-selected four-digit number and will always end up with the number 6,174, which then gets you into an endless loop (i.e. we continuously get back to 6,174).

The consistency of these loops is easy to verify, since there are a finite number of four-digit numbers. Here are some variations for you to ponder — and appreciate! For example, we can also do this with three-digit numbers that are not palindromic. Let's consider the number 759, getting the largest number from these digits would be 957 and the smallest number with these digits would be 579, which when we subtract them, we get 378. Repeating this procedure, we establish the following subtraction $873 - 378 = 485$. Once again repeating the process with the number 485, we get $854 - 458 = 396$. And again, $963 - 369 = 594$. Continuing along, $954 - 459 = 495$. Again, $954 - 459 = 495$, which now gets us into a loop with the number 495. You might have your child consider a test for five-digit numbers using the same scheme. This will once again provide a child with a dimension of investigating mathematics that could nicely augment the regular school curriculum, but with a surprising result that will leave a child in awe!

We should not confuse this Kaprekar constant (6174) with the *Kaprekar numbers*: $9, 45, 297, 703, 4879, \cdots$:, which are more items of interest to exhibit to your child as further demonstration of mathematics' beauty. Here are the amazements of the Kaprekar numbers:

$9^2 = 81,$	$8 + 1 = 9;$
$45^2 = 2025,$	$20 + 25 = 45;$
$297^2 = 88209,$	$88 + 209 = 297;$
$703^2 = 494209,$	$494 + 209 = 703;$
$4879^2 = 23804641,$	$238 + 4641 = 4879;$
$17344^2 = 300814336,$	$3008 + 14336 = 17344;$
$538461^2 = 289940248521,$	$289940 + 248521 = 538461.$

We have just seen how a repetitive process brings us to a common end. In a similar fashion we will enlighten your child with a surprising result. This is about a number that has some truly exceptional properties. We begin by showing how the number 1089 just happens to "pop up" when least expected, and then we'll take another look at this number. We shall begin by having your child select any three-digit number, where the unit and hundreds digit are not the same and follow the instructions below.

Have your child follow these instructions step by step, while we do it along in the boxes below each instruction.

Choose any three-digit number (where the unit and hundreds digit are not the same).

> We will do it with you here — arbitrarily selecting:
>
> **825**

Reverse the digits of this number you have selected.

> We will continue by reversing the digits of 825 to get: **528**

Subtract the two numbers (naturally, the larger minus the smaller)

> Our calculated difference is: **825 − 528 = 297**

Once again, reverse the digits of this difference.

> Reversing the digits of 297 we get the number: **792**

Now, add your last two numbers.

> We then add the last two numbers to get: 297 + 792 = 1089

Your result should be the same[2*] as ours even though your starting number was different from ours.

Your child will probably be astonished that regardless of which number you selected at the beginning, you got the same result as we did, 1089. It might be good to do this again with another number just to make sure that your child realizes that following this process will always end up with number 1089.

[2*] If not, then you made a calculation error. Check it.

How does this happen? Is this a "freak property" of this number? Did we do something devious in our calculations? Of course not! You might have your child be even more impressed with a further property of this lovely number 1089.

Let's look at the first nine multiples of 1089.

$$1089 \times 1 = 1089$$
$$1089 \times 2 = 2178$$
$$1089 \times 3 = 3267$$
$$1089 \times 4 = 4356$$
$$1089 \times 5 = 5445$$
$$1089 \times 6 = 6534$$
$$1089 \times 7 = 7623$$
$$1089 \times 8 = 8712$$
$$1089 \times 9 = 9801$$

Do you notice a pattern among the products? Look at the first and ninth products (i.e. 1089 and 9801). They are the reversals of one another. The second and the eighth products (i.e. 2178 and 8712) are also reversals of one another. And so, the pattern continues, until the fifth product, 5445, is the reverse of itself, known as a palindromic number. Recall that $1089 \times 9 = 9801$, is the reversal of the original number. Amazingly, the same property holds for $10989 \times 9 = 98901$, and similarly, $109989 \times 9 = 989901$.

Your child might be highly impressed to notice that when we altered the original number, 1089, by inserting a 9 in the middle of the number to get 10989, and later extended that by inserting 99 in the middle of the number 1089 to get 109989, the same property held — that is the number reversed. It would be nice to conclude from this that each of the following numbers has the same property: 1099989, 10999989, 109999989, 1099999989, 10999999989, and so on.

As a matter of fact, there is only one other number with four or fewer digits, where a multiple of itself is equal to its reversal, and that is the number 2178 (which just happens to be 2×1089), since $2178 \times 4 = 8712$. Wouldn't it be nice if we could extend this as we did with the above example by inserting 9s into the middle of the number to generate other numbers that have the same property. Yes, it is true that

$$21978 \times 4 = 87912$$
$$219978 \times 4 = 879912$$
$$2199978 \times 4 = 8799912$$
$$21999978 \times 4 = 87999912$$
$$219999978 \times 4 = 879999912$$
$$2199999978 \times 4 = 8799999912$$

and so on.

As if the number 1089 didn't already have enough cute properties, here is another one that (sort of) extends the 1089: We will actually consider the number 1089 in two parts: the numbers 1 and 89. At this point, the parent needs to make a judgment as to whether the child has the patience and interest to investigate this further. You don't want to overkill, yet, if a student seems motivated to investigate this further, we offer that possibility with the following.

This time we shall take any number and get the sum of the squares of its digits. Then we continue this process of finding the sum of the squares of the digits of each succeeding number. Each time, curiously enough, you will eventually reach either 1 or 89. Take a look at some examples that follow.

We will begin with the number 30, and we will find the sum of the digits of this number and then continuously take the sum of the digits of the new resulting numbers:

$$3^2 + 0^2 = 9,$$
$$9^2 = 81,$$
$$8^2 + 1^2 = 65,$$
$$6^2 + 5^2 = 61,$$
$$6^2 + 1^2 = 37,$$
$$3^2 + 7^2 = 58,$$
$$5^2 + 8^2 = \mathbf{89}, \text{ now watch what happens as we continue.}$$
$$8^2 + 9^2 = 145,$$
$$1^2 + 4^2 + 5^2 = 42,$$
$$4^2 + 2^2 = 20,$$

$2^2 + 0^2 = 4,$

$4^2 = 16,$

$1^2 + 6^2 = 37,$

$3^2 + 7^2 = 58,$

$5^2 + 8^2 = \mathbf{89}, \dots$

Once we reached 89 the first time, we got into what we call a *loop*, since we always seem to get back to the number 89, when we repeat the process. Let's try this now with the number 31.

$3^2 + 1^2 = 10,$

$1^2 + 0^2 = \mathbf{1},$

$1^2 = \mathbf{1}$

Again, for the number 1 a loop is formed, getting us back to 1 over and over. To better firm up this concept, we shall now do this for the number 32.

$3^2 + 2^2 = 13,$

$1^2 + 3^2 = 10,$

$1^2 + 0^2 = \mathbf{1},$

$1^2 = \mathbf{1}.$

And now we will use the number 33 to see what happens when we use this process:

$3^2 + 3^2 = 18,$

$1^2 + 8^2 = 65,$

$6^2 + 5^2 = 61,$

$6^2 + 1^2 = 37,$

$3^2 + 7^2 = 58, 5^2 + 8^2 = \mathbf{89}$. From above, you will recall that 89 gets us back to 89.

You might want to have your child try this for other numbers to provide further evidence that this will work for all numbers — another nice way to add some "life" into mathematics!

For the curious secondary school student, who has a grasp of elementary algebra, we will go back to the original oddity of the 1089 — the one where we used digit reversals in order to generate 1089 from a randomly-selected three-digit number — and show *why* this actually works. We assumed that any number we chose would lead us to 1089. How can we be sure? Well, we could try all possible three-digit numbers to see if it works. That would be tedious and not particularly elegant. An investigation of this oddity requires nothing more than some knowledge of elementary algebra. For the curious student we will provide an algebraic explanation as to why it "works."

We shall represent the arbitrarily selected three-digit number, **htu** as $100h + 10t + u$, where h represents the hundreds digit, t represents the tens digit, and u represents the units digit.

Let $h > u^{3*}$, which would be the case either in the number you selected or the reverse of it.

In the subtraction, $u - h < 0$; therefore, take 1 from the tens place (of the minuend) making the units place $10 + u$.

Since the tens digits of the two numbers to be subtracted are equal, and 1 was taken from the tens digit of the minuend, then the value of this digit is $10(t - 1)$. The hundreds digit of the minuend is $h - 1$, because 1 was taken away to enable subtraction in the tens place, making the value of the tens digit $10(t - 1) + 100 = 10(t + 9)$.

We can now do the first subtraction:

$100\,(h - 1)$	$+100\,(t + 9)$	$+(u + 10)$
$100u$	$+10t$	$+h$
$100\,(h - u - 1)$	$+10(9)$	$+u - h + 10$

Reversing the digits of this difference gives us:

$$100\,(u - h + 10) + 10\,(9) + (h - u - 1)$$

Now adding these last two expressions gives us:

$$100\,(9) + 10\,(18) + (10 - 1) = 1089$$

It is important to impress on your child that algebra enables us to inspect the arithmetic process, regardless of the number.

---

[3]*The symbol $>$ means "greater than" and the symbol $<$ means "less than."

Before we leave the number 1089, you might want to point out to the child, who is now so motivated to inspect this curious number further that there is still another oddity, namely, $33^2 = 1089 = 65^2 - 56^2$, which is unique among two-digit numbers. By this time you must agree that there is a particular beauty in the number **1089**. We hope that your child is now becoming hooked on investigating mathematics beyond the school curriculum.

Your child should also be aware that one can have fun with mathematics without actually doing much computation. For example, the construction and the features of *magic squares* can be entertaining and also — in a hidden form — provides an opportunity for quantitative insight and arithmetic practice.

One of the first puzzles in the history of mathematics has been the magic square, and today it is as fascinating as it was ages ago. The task is to find a square arrangement of numbers, so that the sum of the numbers in each row and each column is the same as the sum of the numbers in each of the two diagonals. The oldest known example of a magic square is the Lo Shu square, with numbers arranged on the back of a turtle as shown in Figure 5.1 with dots representing the numbers — shown to the left. It was known to Chinese mathematicians as early as 650 BCE, and became important in Feng Shui, the art of placing objects to achieve harmony with the surrounding environment.

A legend from that time tells us that there was a huge flood on the Lo River in China, and the people tried to placate the river's god. But each time they offered a sacrifice, a turtle emerged from the river, and walked around the offering until a child noticed a strange pattern of dots on the turtle's shell. After studying these markings, the people realized that the correct amount of sacrifices to make would be 15. And after they did so,

Figure 5.1: Lo Shu square and the magic turtle

the river's god was satisfied and the flood receded. The number 15 is the sum of numbers in each row, column, and diagonal of the Lo Shu magic square. We also notice that the middle number of the sequence 1–9 is 5, and is in the middle cell. You might ask your child to notice what relationship there exists among the 3 numbers in each of the line-paths going through the middle 5? They are all in an arithmetic progression, that means that the have a common difference. For example, let's take the vertical line through the center cell, where the numbers are 1, 5 and 9. Their common difference is 4. As another example, the horizontal line through the center cell we have the numbers 3, 5 and 7. Their common difference is 2. By now your child should be able to recognize the others as well. Furthermore, you might guide your child to realize the sums of the squares of the first and third columns are equal, namely, $4^2 + 3^2 + 8^2 = 2^2 + 7^2 + 6^2 = 89$. You might want to encourage a child to seek other possible number patterns from this magic square. For example, the sum of the squares of the middle column $9^2 + 5^2 + 1^2 = 107 = 89 + 18$. Now you may wonder what this could relate to? If you consider the sums of the squares of the numbers in the 3 horizontal rows, we find that they are 101, 83 and 101. Now if we take $101 - 83$, we get the 18, which we encountered earlier.

As we said before, these magic squares offer us a good bit of entertainment, which, as parents, you should introduce your child to these wonders. It is rare — and unfortunate — that their teacher will take time from the standard curriculum to do these entertainments, however, they are essential in generating a positive feeling towards mathematics. So, consider then another relationship on this magic square. Let us consider the three rows of digits as numbers, and then take the sum of the squares: $492^2 + 357^2 + 816^2 = 1,035,369$. Now we will take the reversals of each of these numbers, and find the sum of the squares: $294^2 + 753^2 + 618^2 = 1,035,369$. Lo and behold, the sums of the same. This is truly amazing, and therefore, it is called a magic square. Incidentally, the same is true for the columns, as we can see from the following computation: $438^2 + 951^2 + 276^2 = 1,172,421$, and $834^2 + 159^2 + 672^2 = 1,172,421$. There can be other patterns to be found of the similar nature involving the diagonals, we leave that for your child to discover.

There is one magic square, however, that stands out from among the rest for its beauty and additional properties — not to mention its curious appearance. This particular magic square has many properties beyond those required for a square arrangement of numbers to be considered "magic." This magic square even comes to us through art, and not through the usual mathematical channels. It is depicted in the background of the famous 1514 engraving by the renowned German artist Albrecht Dürer (1471–1528), who lived in Nürnberg, Germany. (See Figure 5.2)

Remember, a magic square is a square arrangement of numbers, where the sum of the numbers in each of its columns, rows, and diagonals is the same. As we begin to examine the magic square in Dürer's etching, we

Figure 5.2: *Melencolia I*, engraving, Albrecht Dürer (1514)

Figure 5.3:   Initials AD of Albrecht Dürer and the year 1514

Figure 5.4:   Dürer's magic square

should take note that most of Dürer's works were signed by him with his initials, one over the other, and with the year in which the work was made included. Here we find it in the dark-shaded region near the lower right side of the picture (Figures 5.2 and 5.3). We notice that it was made in the year 1514.

The observant reader may notice that the two center cells of the bottom row of the Dürer magic square also depicts the year the etching was made. Let us examine this magic square more closely. (See Figure 5.4.)

First let's make sure that it is, in fact, a true magic square, by checking to see if the sum of each of the rows, columns and diagonals have the same sum. In this case, the sum is 34. This is all that would be required for this square matrix of numbers to be considered a "magic square." However, this "Dürer Magic Square" has lots more properties that other magic squares do not have. Let us now marvel about some of these extra properties. In practice, parents might be wise to allow their children to discover some of these number relationships before exposing them. Hints may be provided to get them started in their search.

- The four corner numbers have a sum of 34:
  $16 + 13 + 1 + 4 = 34$
- Each of the four corner 2 by 2 squares has a sum of 34:
  $16 + 3 + 5 + 10 = 34$
  $2 + 13 + 11 + 8 = 34$
  $9 + 6 + 4 + 15 = 34$
  $7 + 12 + 14 + 1 = 34$
- The center 2 by 2 square has a sum of 34.
  $10 + 11 + 6 + 7 = 34$
- The sum of the numbers in the diagonal cells equals the sum of the numbers in the cells not in the diagonal.
  $16+10+7+1+4+6+11+13 = 3+2+8+12+14+15+9+5 = 68$
- The sum of the squares of the numbers in both diagonal cells is
  $16^2 + 10^2 + 7^2 + 1^2 + 4^2 + 6^2 + 11^2 + 13^2 = 748$
  This number is equal to

  o the sum of the squares of the numbers not in the diagonal cells:
  $3^2 + 2^2 + 8^2 + 12^2 + 14^2 + 15^2 + 9^2 + 5^2 = 748$
  o the sum of the squares of the numbers in the first and third rows:
  $16^2 + 3^2 + 2^2 + 13^2 + 9^2 + 6^2 + 7^2 + 12^2 = 748$
  o the sum of the squares of the numbers in the second and fourth rows.
  $5^2 + 10^2 + 11^2 + 8^2 + 4^2 + 15^2 + 14^2 + 1^2 = 748$
  o the sum of the squares of the numbers in the first and third columns.
  $16^2 + 5^2 + 9^2 + 4^2 + 2^2 + 11^2 + 7^2 + 14^2 = 748$
  o the sum of the squares of the numbers in the second and fourth columns.
  $3^2 + 10^2 + 6^2 + 15^2 + 13^2 + 8^2 + 12^2 + 1^2 = 748$

- The sum of the cubes of the numbers in the diagonal cells equals the sum of the cubes of the numbers not in the diagonal cells.
$$16^3 + 10^3 + 7^3 + 1^3 + 4^3 + 6^3 + 11^3 + 13^3 =$$
$$= 3^3 + 2^3 + 8^3 + 12^3 + 14^3 + 15^3 + 9^3 + 5^3 = 9,248$$
- Notice the following beautiful symmetries:
$$2 + 8 + 9 + 15 = 3 + 5 + 12 + 14 = 34$$
$$2^2 + 8^2 + 9^2 + 15^2 = 3^2 + 5^2 + 12^2 + 14^2 = 374$$
$$2^3 + 8^3 + 9^3 + 15^3 = 3^3 + 5^3 + 12^3 + 14^3 = 4624$$
- Adding the first row to the second, and the third row to the fourth, produces a pleasing symmetry:

$16 + 5 = 21$	$3 + 10 = 13$	$2 + 11 = 13$	$13 + 8 = 21$
$9 + 4 = 13$	$6 + 15 = 21$	$7 + 14 = 21$	$12 + 1 = 13$

Adding the first column to the second, and the third column to the fourth, produces a pleasing symmetry:

$16 + 3 = 19$	$2 + 13 = 15$
$5 + 10 = 15$	$11 + 8 = 19$
$9 + 6 = 15$	$7 + 12 = 19$
$4 + 15 = 19$	$14 + 1 = 15$

A nice activity would be to join your child to search for other patterns in this beautiful magic square. Remember, this is not a typical magic square, where all that would be required is that all the rows, columns and diagonals have the same sum. This Dürer magic square has many more properties.

Why is $x \cdot 0 = 0$?

Most people will not remember how they felt when they heard that $x \cdot 0 = 0$ for the first time. Consequently, they find it very difficult to explain this apparently elementary fact to others, especially to their children, who are often told in school to learn it and accept it as truth. Of course, this is bound to create an uneasy feeling idea of having missed out on an explanation: "If I hold a dollar in my hand and multiply it by zero, is it really gone? Who took it?".

One has to be aware that — when encountered for the first time — the fact $x \cdot 0 = 0$ might appear unusual and strange and that it takes some time for a child to grasp its meaning. When we try to come up with explanations, we should, as always, be patient and appeal to the child's basic understanding. And, most importantly, while talking to the child, one should try to figure out the sources of possible misunderstandings. First, make sure the child knows that "0" is a symbol meaning the absence of something or simply means "nothing." Doing something zero times means not doing it at all. Next, the child must have a basic understanding of products, like 6 times 3 means to take six groups of three, or equivalently, three groups of six. If one takes zero groups of six means that one takes no group at all, and that leaves one with nothing, that is, "0." The same result is obtained if one takes six groups of zero, meaning that one takes six times nothing and that remains nothing.

This might still be difficult to grasp for a child, who is not used to the way of speaking like "take zero groups of something". It is important to know in which ways this could be misunderstood and to be able to offer alternative explanations. One might be asked "If I have 6 and add it three times, I get $6 + 6 + 6 = 6 \times 3$. But if I add it zero times, I do nothing at all to the original 6 and therefore I'm still left with the group of 6. So why isn't $6 \times 0 = 6$?" Here we see that the problem is probably one of rhetoric. It has more to do with our way of speaking, than with mathematics: Doing nothing, shouldn't that leave you with what you have? But what do you "have" to start with? 6×0 just means that you "have" zero groups of 6, that is, nothing. Thus, multiplying by 0 is not equivalent with "doing nothing" (which would leave 6 as it is).

And sometimes it helps to change the picture. Think of 6 times 3 as taking 6 steps of length 3. If you take a step of length 0 you remain where you are. You can take as many steps of length zero as you want without moving forward. Or, to the same effect, you can take zero steps of an arbitrary length, that is, no step at all, and again you will not change your position. Zero could be a difficult concept for your child to grasp but further discussion about zero might provide some insight.

Why Can't We Divide by Zero?

Beyond the famous Ten Commandments, we in mathematics have an 11th commandment, that is, *thou shalt not divide by zero!* One would hope that most teachers early on would not only mention that to their class, but also demonstrate what happens if you are permitted to divide by zero. There are a number of demonstrations that lead to absurdities when dividing by zero. Here is where the parent can provide a closing of an omission from the school program. So, let us embark on an inspection of this unusual number zero.

Your child might well ask, why is division by zero not permissible? We in mathematics pride ourselves in the order and beauty in which everything in the realm of mathematics falls neatly into place. When something arises that could spoil that order, we simply *define* it to suit our needs. This is precisely what happens with division by zero. So, let's give this "commandment" some meaning.

Consider the quotient $\frac{n}{0}$. Without acknowledging the division-by-zero commandment, let us speculate (i.e. guess) what the quotient might be. Let us say it is p. In that case, we could check by multiplying $0 \times p$ to see if it equals n as would have to be the case for the division to be correct. We know that there is no number a, where $0 \times p \neq a$, since $0 \times p = 0$. So there is no number, p that can be the quotient to this division. For that reason, we define division by zero to be invalid.

A more convincing case for defining away division by zero is to show students how it can lead to a contradiction of an accepted fact, namely, that obviously $1 \neq 2$. We will show them that were division by zero acceptable, then $1 = 2$, clearly an absurdity!

Here is the "proof" that $1 = 2$:

$$\text{Let } a = b$$
$$\text{Then } a^2 = ab \qquad [\text{multiplying both sides by } a]$$
$$a^2 - b^2 = ab - b^2 \qquad [\text{substracting } b^2 \text{ from both sides}]$$
$$(a-b)(a+b) = b(a-b) \qquad [\text{Factoring}]$$
$$a + b = b \qquad [\text{Dividing by } (a-b)]$$
$$2b = b \qquad [\text{replace } a \text{ by } b]$$
$$2 = 1 \qquad [\text{Divide both sides by } b]$$

Your child may now be a bit baffled by this seeming-correct result, which is truly an absurdity. Have him or her consider what happened in the step where we divided by $(a - b)$. Looking back at the values at the beginning, we actually divided by zero, because $a = b$, so $a - b = 0$. That ultimately led us to an absurd result, leaving us with no option other than to prohibit division by zero. By taking the time to explain this rule about division by zero your child will be getting a very important lesson from a parent — perhaps even one better than that provided by the teacher. It is unfortunate, that too many teachers just *tell* students they are not allowed to divide by zero without any explanation, such as we have provided here so that they will have a much better appreciation for, and understanding of, mathematics.

If you want to drive this point home a bit further with some entertainment, try this logical reasoning and you will firm up the notion of not being allowed to divide by zero. Mentioned to your child that we can "prove" that $1 = 2$ by using the forbidden division by zero. We know that if $5a = 5b$ then $a = b$. However, if we use the reasoning that since $1 \times 0 = 0$, and $2 \times 0 = 0$, we can conclude that $1 \times 0 = 2 \times 0$. Then, using the same reasoning as before, we can say that $1 = 2$, by dividing both sides by zero. Since such absurdities result from division by zero, it is forbidden in mathematics usage. Remember, the eleventh commandment: Thou shall not divide by zero.

When Cancellations are Mistakes — and When Not!

Parents can also entertain their child with some odd and incorrect mathematics that strangely yields a correct answer. Here is one that has been sometimes referred to as a "howler." It is just the kind of mistake that can surely bring us to wonder! Consider the following, where we are asked to reduce the fraction $\frac{16}{64}$, and we simply cross out the 6's: $\frac{16}{64} = \frac{1\cancel{6}}{\cancel{6}4} = \frac{1}{4}$, which, strangely enough, gives us the correct result. We can apply this method to the following fractions as well:

To reduce the fraction $\frac{26}{65}$, we simply cross out the 6s to get the right answer:

$$\frac{26}{65} = \frac{2\cancel{6}}{\cancel{6}5} = \frac{2}{5}.$$

To reduce the fraction $\frac{19}{95}$, we simply cross out the 9s to get the right answer:

$$\frac{19}{95} = \frac{1\cancel{9}}{\cancel{9}5} = \frac{1}{5}.$$

To reduce the fraction $\frac{49}{98}$, we simply cross out the 9s to get the right answer

$$\frac{49}{98} = \frac{4\cancel{9}}{\cancel{9}8} = \frac{4}{8} \left(= \frac{1}{2} \right).$$

Naturally, this can be done for all two-digit multiples of 11 $\left(\frac{1\cancel{1}}{\cancel{1}1}, \frac{2\cancel{2}}{\cancel{2}2}, \ldots \right)$, but that is as far as we can apply this silly method for two-digit numbers. One then wonders why this simple (or silly) method cannot always be used.

Sometimes an erroneous method can lead (just by coincidence) to a correct result, as in the cases above. The danger, of course, is that we must not generalize this method. It is very important that the parent emphasize that these are the only two-digit examples of this mistaken procedure that leads to a correct reduction of a fraction.

Parents should use proper judgment as to when to take this discussion further as it requires some sophistication. An arithmetic explanation for why this works can be seen from the following calculation:

$$\frac{16}{64} = \frac{1 \times 10 + 6}{10 \times 6 + 4} = \frac{\cancel{6} \times \frac{16}{6}}{\cancel{6} \times \frac{64}{6}} = \frac{\cancel{6} \times \frac{8}{3}}{\cancel{6} \times \frac{32}{3}} = \frac{8}{32} = \frac{1}{4}.$$

For those readers with a good working knowledge of elementary algebra, we can "explain" this situation, and show that the four fractions above are the *only* ones (composed of two-digit numbers) where this type of cancellation will hold true. (This explanation uses only elementary algebra.)

We begin by considering the fraction $\frac{10x + a}{10a + y}$.

The above four cancellations were such that when canceling the a's the fraction was equal to $\frac{x}{y}$.

Therefore, $\dfrac{10x + a}{10a + y} = \dfrac{x}{y}$

This yields: $y(10x + a) = x(10a + y)$

$10xy + ay = 10ax + xy$

$9xy + ay = 10ax$

And so $y = \dfrac{10ax}{9x + a}.$

At this point we shall inspect this equation. It is necessary that x, y and a are integers, since they were digits in the numerator and denominator of a fraction. It is now our task to find the values of a and x for which y will also be integral.

To avoid a lot of algebraic manipulation, you will want to set up a chart, which will generate values of y from $y = \frac{10ax}{9x+a}$. Remember that x, y and a must be single digit integers. Below is a portion of the table you will be constructing. Notice that the cases where $x = a$ are excluded since $\frac{x}{a} = 1$.

x/a	1	2	3	4	5	6	7	8	9
1		$\frac{20}{11}$	$\frac{30}{12}$	$\frac{40}{13}$	$\frac{50}{14}$	$\frac{60}{15}=4$	$\frac{70}{16}$	$\frac{80}{17}$	$\frac{90}{18}=5$
2	$\frac{20}{19}$		$\frac{60}{21}$	$\frac{80}{22}$	$\frac{100}{23}$	$\frac{120}{24}=5$	$\frac{140}{25}$	$\frac{160}{26}$	$\frac{180}{27}$
3	$\frac{30}{28}$	$\frac{60}{29}$		$\frac{120}{31}$	$\frac{150}{32}$	$\frac{180}{33}$	$\frac{210}{34}$	$\frac{240}{35}$	$\frac{270}{36}$
4	$\frac{40}{37}$	$\frac{80}{38}$	$\frac{120}{39}$		$\frac{200}{41}$	$\frac{240}{42}$	$\frac{280}{43}$	$\frac{320}{44}$	$\frac{360}{45}=8$
5	$\frac{50}{46}$	$\frac{100}{47}$	$\frac{150}{48}$	$\frac{200}{49}$		$\frac{300}{51}$	$\frac{350}{52}$	$\frac{400}{53}$	$\frac{450}{54}$
6	$\frac{60}{55}$	$\frac{120}{56}$	$\frac{180}{57}$	$\frac{240}{58}$	$\frac{300}{59}$		$\frac{420}{61}$	$\frac{480}{62}$	$\frac{540}{63}$
7	$\frac{70}{64}$	$\frac{140}{65}$	$\frac{210}{66}$	$\frac{280}{67}$	$\frac{350}{68}$	$\frac{420}{69}$		$\frac{560}{71}$	$\frac{630}{72}$
8	$\frac{80}{73}$	$\frac{160}{74}$	$\frac{240}{75}$	$\frac{320}{76}$	$\frac{400}{77}$	$\frac{480}{78}$	$\frac{560}{79}$		$\frac{720}{81}$
9	$\frac{90}{82}$	$\frac{180}{83}$	$\frac{270}{84}$	$\frac{360}{85}$	$\frac{450}{86}$	$\frac{540}{87}$	$\frac{630}{88}$	$\frac{720}{89}$	

The portion of the chart, pictured above, already generated two of the four integral values of y; that is, when $x = 1$, and $a = 6$, then $y = 4$, and when $x = 2$, and $a = 6$ then $y = 5$. These values yield the fractions $\frac{16}{64}$ and $\frac{26}{65}$, respectively. The remaining two integral values of y will be obtained when $x = 1$, and $a = 9$, yielding $y = 5$, and when $x = 4$, and $a = 9$, yielding $y = 8$. These generate the fractions $\frac{19}{95}$ and $\frac{49}{98}$, respectively. This should convince you that there are only four such fractions composed of two-digit numbers.

Your child may wonder if there are fractions composed of numerators and denominators of more than two digits, where this strange type of cancellation holds true. Following are some examples of three-digit numbers where this sort of strange cancellation can work.

$$\frac{199}{995} = \frac{19\cancel{9}}{\cancel{9}95}\left(=\frac{1}{5}\right), \quad \frac{266}{665} = \frac{26\cancel{6}}{\cancel{6}65}\left(=\frac{2}{5}\right), \quad \frac{124}{217} = \frac{\cancel{1}24}{2\cancel{1}7}\left(=\frac{4}{7}\right),$$

$$\frac{103}{206} = \frac{1\cancel{0}3}{2\cancel{0}6} = \frac{13}{26}\left(=\frac{1}{2}\right), \quad \frac{495}{990} = \frac{49\cancel{5}}{99\cancel{0}} = \frac{45}{90}\left(=\frac{1}{2}\right),$$

$$\frac{165}{660} = \frac{16\cancel{5}}{66\cancel{0}} = \frac{15}{60}\left(=\frac{1}{4}\right), \quad \frac{127}{762} = \frac{12\cancel{7}}{76\cancel{2}}\left(=\frac{1}{6}\right), \quad \text{and}$$

$$\frac{143185}{1701856} = \frac{143\cancel{1}85}{170\cancel{1}856} = \frac{1435}{17056}\left(=\frac{35}{416}\right)$$

Merely to have some fun with mathematics, we offer here some further peculiarities of this sort. Have your child consider reducing the following fraction: $\frac{499}{998}$. He or she should find that $\frac{499}{998} = \frac{4}{8} = \frac{1}{2}$. This can be extended as follows:

$$\frac{19999}{99995} = \frac{1999\cancel{9}}{\cancel{9}9995} = \frac{199\cancel{9}}{\cancel{9}995} = \frac{19\cancel{9}}{\cancel{9}95} = \frac{1\cancel{9}}{\cancel{9}5} = \frac{1}{5},$$

$$\text{that is,} \quad \frac{19999}{99995} = \frac{1\cancel{9}\cancel{9}\cancel{9}\cancel{9}}{\cancel{9}\cancel{9}\cancel{9}\cancel{9}5} = \frac{1}{5}, \text{ or}$$

$$\frac{26666}{66665} = \frac{2666\cancel{6}}{\cancel{6}6665} = \frac{266\cancel{6}}{\cancel{6}665} = \frac{26\cancel{6}}{\cancel{6}65} = \frac{2\cancel{6}}{\cancel{6}5} = \frac{2}{5},$$

$$\text{that is,} \quad \frac{26666}{66665} = \frac{2\cancel{6}\cancel{6}\cancel{6}\cancel{6}}{\cancel{6}\cancel{6}\cancel{6}\cancel{6}5} = \frac{2}{5}$$

A pattern is now emerging and you may realize that

$$\frac{49}{98} = \frac{499}{998} = \frac{4999}{9998} = \frac{49999}{99998} = \cdots$$

$$\frac{19}{95} = \frac{199}{995} = \frac{1999}{9995} = \frac{19999}{99995} = \frac{199999}{999995} = \cdots$$

$$\frac{26}{65} = \frac{266}{665} = \frac{2666}{6665} = \frac{26666}{66665} = \frac{266666}{666665} = \cdots$$

Enthusiastic children may wish to justify these extensions of the original fractions. Parents who, at this point, have a further desire to seek out additional fractions that permit this strange cancellation should consider the following fractions. Your child may wish to verify the legitimacy of this strange cancellation and then set out to discover more such fractions.

$$\frac{3\not{3}2}{8\not{3}0} = \frac{32}{80} = \frac{2}{5}$$

$$\frac{3\not{8}5}{8\not{8}0} = \frac{35}{80} = \frac{7}{16}$$

$$\frac{1\not{3}8}{\not{3}45} = \frac{18}{45} = \frac{2}{5}$$

$$\frac{2\not{7}5}{7\not{7}0} = \frac{25}{70} = \frac{5}{14}$$

$$\frac{1\not{6}\not{3}}{\not{3}2\not{6}} = \frac{1}{2}$$

However, be careful, because $\frac{16\not{3}}{\not{3}26} \neq \frac{1}{2}$ and $\frac{1\not{6}3}{32\not{6}} \neq \frac{1}{2}$.

Parents should realize that aside from providing an algebraic application to an arithmetic issue, which can be used to introduce a number of important topics in a motivational way, this topic can also provide some recreational activities.

As we continue our journey through the fun aspects of mathematics, we should consider some fun with clever problem-solving techniques, which can surely entertain and motivate your child towards a greater appreciation for mathematics. We can begin with a very simple problem, one that parents can easily explain to their child and be sure to allow them enough time to think about the procedure to the solution. They are likely to give the wrong answer, as the problem could be considered somewhat "tricky."

Consider the following: suppose there is a 100-foot-deep well, and a frog at the bottom of the well. In an attempt to get out of the well, the frog uses all his effort every morning and climbs up 3 feet. Unfortunately, every afternoon he slides down 2 feet. The question is how many days will it take him to reach the top of the well?

Most kids' "knee-jerk reaction" will be to say one hundred days. They see the frog's progress of 1 foot per day, and so they assume that would take 100 days to get out of 100-foot-deep well. However, with a little prompting from the parent, your child should realize that after 97 days, clearly, 97 feet have been achieved. On the morning of the 98th day, the frog will climb 3 feet higher, which will bring him to the top of the well. Therefore, they will see that in $97\frac{1}{2}$ days the frog will have climbed out of the well. At first glance, one might feel that perhaps there is more entertainment than mathematics here. Yet, your child should realize that this kind of thinking is an integral part of the kind of thinking we do in solving mathematical problems. Simple, yet time worthy!

One of the things that many parents would like to do is to have their children begin to think in an original fashion especially in a problem-solving situation. There are many problem-solving techniques that students pick up over the years usually by experience. For example, while grappling with a problem one could be expected to "think out of the box." Essentially, what is being suggested is to avoid trying to solve a problem in the traditional and expected fashion, but to look at the problem from a different point of view. Practically, by definition, this could be considered a counterintuitive way of thinking. This can even be true when a rather simple problem is posed and the straightforward solution could get to be a bit complicated. One might say that many people look at a problem in a psychologically traditional way: the way it is presented and played out. To illustrate this point, we offer a problem here to convince you of an alternate way of thinking. Try the problem yourself (don't look below at the solution) and see whether you fall into the "majority-solvers" group. The solution offered later will probably enchant (as well as provide future guidance to) most children. Thus, presenting this to your child could have a lasting effect on how he or she will more critically approach a problem-solving situation. Needless to say, it should also have a favorable effect on standardized test results.

> *The problem:* **A single elimination (one loss and the team is eliminated) basketball tournament has 25 teams competing. How many games must be played until there is a single tournament champion?**

Typically, and surely not the most elegant procedure is the common way to approach this problem, which is to simulate the tournament, by selecting

24 teams to play in the first round (with one team drawing a bye). This will eliminate 12 teams (12 games have now been played). Similarly, of the remaining 13 teams, 6 play against another 6, leaving 7 teams in the tournament (18 games have been played now). In the next round, of the 7 remaining teams, 3 can be eliminated (21 games have so far been played). The four remaining teams play and eliminate 2 teams, leaving 2 teams for the championship game (23 games have now been played). This championship game is the 24th game.

A much simpler and more sophisticated way to solve this problem, one that most people do not naturally come up with as a first attempt, is to focus only on the losers and not on the winners as we have done above. We then ask the key question: "How many losers must there be in the tournament with 25 teams in order to have one winner?" The answer is simple: 24 losers. How many games must be played to get 24 losers? Naturally, 24. So there you have the answer, very simply done. Now most people will ask themselves, "Why didn't I think of that?" The answer is, it was contrary to the type of training and experience we have had. Becoming aware of the strategy of looking at the problem from a different point of view may sometimes reap nice benefits, as was the case here. Parents should look for other problems of this nature to drive home this important point.

After your child feels comfortable about this alternative technique with solving the problem, you might have him or her consider yet, another way of looking at a solution to this problem — albeit quite similar to the previous solution. That is, to create an artificial situation where out of the 25 teams, we will make 24 of these teams, high-school-level players, and the 25th team a professional basketball team, such as the New York Knickerbockers, which we will assume is superior to all of the other teams, and will easily defeat each one. In this artificially construed situation we would have each of the 24 high school teams playing the Knickerbockers, and as expected lose the game. Hence, after 24 games a champion (in this case, the New York Knickerbockers) is achieved.

Parents are often concerned about how they can improve their child's problem-solving skills. This can comes from being exposed to techniques that may be other than those that come naturally to some students. It sometimes takes an unusual solution to a simple mathematics problem to reinforce clever thinking procedure. One such strategy is sometimes used subconsciously by many people. We are referring to making a decision based on using extremes. We often use extremes camouflaged in the phrase "the worst-case scenario." This approach might be considered as we decide which way to pursue an issue. Extremes can be considered on both the positive and the negative side of a given situation. Such a use of an extreme generally brings us to a good decision. We will present a few such situations here so that parents can use to get their child accustomed to the kind of thinking that can be a useful mathematics problem-solving technique.

Let's begin by considering the windshield of a car that appears to get wetter the faster a car is moving in a rainstorm. This could lead some people to conclude that the car would not get as wet if it were to move slower. A next question that can be asked is it better to walk slowly in a rainstorm, or to run fast, so that you can minimize your wetness? Setting aside the amount of wetness that the front of your body might get from the storm, let us consider two extreme cases for the top of the head: first, going infinitely fast, and second, going so slowly as to practically be stationary. In the first case, there will be a certain amount of wetness on the top of the head. But, if we proceed at practically a 0-mph speed, we would get drenched! Therefore, we conclude that the faster you move, the dryer you stay. This and other such examples should be discussed with your child in a rather calm and conversational style.

The previous illustration of this rather useful problem-solving technique, *using extreme cases*, demonstrates how we use this strategy to clearly sort out an otherwise cumbersome problem situation. We also use this same strategy of considering extremes in more everyday situations. A person who plans to buy an item where bargaining plays a part, such as buying a house or an item at a garage sale, must determine a strategy to make the seller an offer. One must decide what the lowest (extreme) price ought to be and what the highest (extreme) price might be, and then orient him or herself from there. In general, we often consider the extreme values of anything we plan to purchase and then make our decision about which price to settle on based on the extreme situations.

Extreme cases are also utilized when we seek to test a product, say stereo speakers. We would want to test them at extremely low volume and at extremely high volume. We would then take for granted (with a modicum of justification), that speakers that pass the extreme-conditions test would also function properly between these extreme situations.

For parents of somewhat older children or particularly clever children, you might want to expose them to a very difficult problem that is made very very simple with the technique of using extremes. Without using the technique of extremes the following problem would even be very challenging for mathematicians.

A word about the problem before we present it. The beauty of the problem is the elegant solution that involves the use of considering an extreme. After you pose the problem to your child, and he or she considers a method of solution, allow time to perhaps struggle a bit before giving up (if need be), then we will consider the elegant solution provided here that is based on using an extreme situation. Here is the problem:

> **We have two one-gallon bottles. One bottle contains a quart of red wine and the other bottle contains a quart of white wine. We take a tablespoonful of red wine and pour it into the white wine. Then we take a tablespoon of this new mixture (white wine and red wine) and pour it into the bottle of red wine. Is there more red wine in the white-wine bottle, or more white wine in the red-wine bottle?**

To solve the problem, we can figure this out in any of the usual ways — some are often referred to in the high school context as "mixture problems" — or we can use some clever logical reasoning and look at the problem's solution as follows: With the first "transport" of wine there is only red wine on the tablespoon. On the second "transport" of wine, there is as much white wine on the spoon as there remains red wine in the "white-wine bottle." This may require some abstract thinking a bit, but some children should "get it" soon.

The simplest solution to understand, and the one that demonstrates a very powerful strategy is that of *using extremes*. Let us now employ this strategy for this problem. To do this, we will consider the tablespoonful quantity to be a bit larger. Clearly the outcome of this problem is independent of the quantity transported. Therefore, let use an *extremely* large quantity. We will let this quantity actually be the *entire* one quart — the extreme amount. Following the instructions given in the problem statement, we will take this entire amount (one quart of red wine), and pour it into the white-wine bottle. This mixture is now 50% white wine and 50% red wine. We then pour one quart of this mixture back into the red-wine bottle. The mixture is now the same in both bottles. Therefore, we can conclude that there is as much white wine in the red-wine bottle as there is red wine in the white-wine bottle, and the problem is solved!

Once your child has truly understood the solution you might have him or her consider another form of an extreme case, where the spoon doing the wine transporting has a zero quantity. In this case, the conclusion follows immediately: There is as much red wine in the white-wine bottle (none) as there is white wine in the red-wine bottle (none). Once again by using extremes we very easily solved the problem in a rather elegant fashion.

Throughout this chapter, we have proposed a wide variety of ways that parents can add some levity to mathematics learning. There is much to be gained here. First, we hope that the parent will begin to enhance an appreciation for mathematics beyond the school curriculum and then pass on this newly developed appreciation to their children. This can be done through specific examples provided here and others analogous to those we offer. It can also be done through a motivational enthusiasm that hopefully would be contagious to their children. In short, enriching mathematics instruction entertaining aspects such as these presented in this chapter can serve to generate more enthusiastic mathematics learners amongst today's students.

Chapter 6

Some Entertaining Math Stories

As we conclude this book to prepare parents to be supported in the math instruction of their children, we felt that interesting stories from the history of mathematics would be appropriate to share, so that parents can entertain their children with such stories in leisure time, such as during a drive or at dinner to show that math can be entertaining from a historical point of view. Each of these little stories will have a clear message; sometimes, a human nature point of view and other times clever mathematics attached.

The French Mathematician Blaise Pascal

Here is a story you can tell your child at various ages. It is about an unusually gifted person who broke ground in various areas of mathematics. One of the greatest mathematicians of all time was Blaise Pascal who was born in Clermont, Auvergne, France on June 19, 1623.

Blaise Pascal

His father Etienne Pascal was a politician and a man of culture and intellectual distinction. Blaise Pascal's mother died when he was four years old and was, thus, raised along with his two sisters by his father. His early years, encouraged by his father, found him deeply engaged in religious thinking. This often distracted him from other intellectual endeavors. When Pascal was seven years old, the father and his three children moved to Paris. This was about the time when the father was heavily involved in teaching his children at home. Pascal was not physically well conditioned, yet this was compensated by an exceptionally brilliant mind. Pascal's father was impressed at how quickly his young son would pick up new ideas of what was then considered the classical education. He kept mathematics at a distance from him, so as not to put too much strain on the young child. Actually, this built up Pascal's curiosity about mathematics even more. Once the father realized his son's incredible mathematics talents, he gave him a copy of Euclid's *Elements*, perhaps one of the first compilations of a logical development of geometry and other aspects of mathematics. As a matter of fact, this book is the basis for the American high school geometry course and much more. Pascal's sister claimed that her younger brother had discovered Euclid's first 32 propositions in the same order in which Euclid did, yet, without referring to the book. It was the 32nd proposition that the sum of the angles of a triangle is equal to the sum of two right angles (i.e. 180°) that further demonstrated Pascal's unique talent.

By the age of 14, Pascal was admitted to the weekly meetings of a group that eventually developed into the French Academy of Science. About the time when he was 16 years old, and having been motivated by the work of Girard Desargues (1591–1661), he got involved in geometry and proved some of the most beautiful theorems in geometry, some of which today bear his name (Pascal's Theorem) We show one of them as it is very easy to demonstrate to your child by allowing him or her to actually perform this "experiment." All you need is a circle and a ruler. We show this configuration in Figure 6.1, where we select six randomly selected points on a circle. Then we join these points consecutively, forming a hexagon, which is inscribed in the circle (of course, avoid having any pair of opposite sides parallel). We extend the pairs of opposite sides (*AB-ED, BC-FE,* and *CD*-AF) so that the lines they intersect. As shall mark these points of intersection, *L*, *N* and *M*, and find that these points will always lie on a straight line. This would be

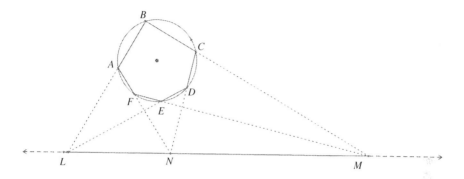

Figure 6.1: Pascal's Theorem

true for any six points on a circle, and curiously enough, this can also be extended to selecting any six points on an ellipse. At first other famous mathematicians of the times, such as René Descartes, refused to believe that such a discovery could be made by a 16-year-old boy. But in time, it was properly accepted.

Pascal had to pay a price for his brilliance. From the age of 17 until the end of his life at age 39, he lived in physical pain with sleepless nights and unpleasant days. Yet, he kept on working. At age 19 Pascal invented the first calculator machine (See Figure 6.2) in order to assist his father's computational work as a tax collector for the city of Rouen. This calculator was able to do addition and subtraction and was referred to as Pascal's calculator or *Pascaline*. There are currently four versions of this machine exhibited in the <u>Musée des Arts et Métiers</u> in Paris. At the time of its development, the machine was considered a luxury item and this motivated Pascal to continue to improve its functioning over the next 10 years.

The society in which he lived was tormented by religious upheaval, which to some extent affected Pascal as well, as his dear sister Jacqueline, who had supported him, entered a monastery in Pert-Royal and left him to live alone, something which he was not accustomed to. At age 23 he suffered a temporary paralysis, but his intellectuality continued unabated. He continued to lead a rather turbulent life tortured by his family's involvement in various religious followings. In 1654 at age 31, Pascal engaged in probably the most important contribution he had made to mathematics. That is, he embarked on a mathematical correspondence with Pierre de Fermat, which

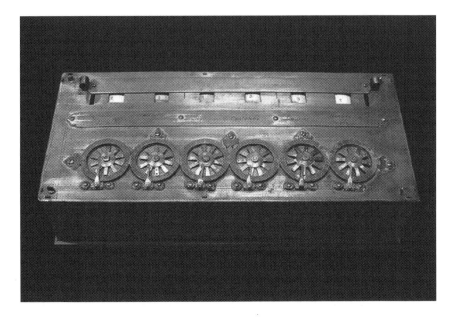

Figure 6.2: Pascal's Calculating Machine

```
                                        1    .    .    .    .    .    .    .    .    .    2⁰
                                   1         1    .    .    .    .    .    .    .    .    2¹
                              1         2         1    .    .    .    .    .    .    .    2²
                         1         3         3         1    .    .    .    .    .    .    2³
                    1         4         6         4         1    .    .    .    .    .    2⁴
               1         5        10        10         5         1    .    .    .    .    2⁵
          1         6        15        20        15         6         1    .    .    .    2⁶
     1         7        21        35        35        21         7         1    .    .    2⁷
1         8        28        56        70        56        28         8         1    .    2⁸
1    9        36        84       126       126        84        36         9         1    . 2⁹
1   10    45       120       210       252       210       120        45        10    1  2¹⁰
```

Figure 6.3: The Pascal Triangle Showing Powers of 2

eventually became the basis for the theory of probability. During the year 1654, Pascal and Fermat challenged each other with mathematical problems that began to generate a thinking in probability, as we know it today. One of the early problems posed involved a game where two players would gain points, with a specified number of points to win the game.

Throughout this time Pascal made considerable use of the triangular arrangement of numbers that also bears his name today. In Figure 6.3 we see this arrangement, where beginning at the top with a 1, followed by a

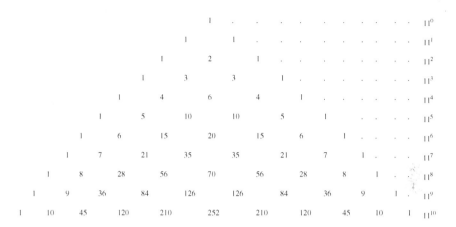

Figure 6.4: The Pascal Triangle Showing Powers of 11

second row of two 1s, and each succeeding row begins with a 1 on each end, and each other number between the 1s is the sum of the two numbers above it on either side of it. This arrangement of numbers is known today as the Pascal triangle. Curiously, many number relationships can be found on the Pascal triangle. For example, the sum of the numbers in each row is a power of 2 as shown in Figure 6.3.

When we look at Figure 6.4, we also notice that we have a representation of powers of 11. This triangular arrangement of numbers is very useful to this day when working with probability.

It could be said, that Pascal's name in today's recollection of the history of mathematics is being a co-inventor of the theory of probability, which seems to become increasingly more important in our everyday lives from weather prediction to work in finance. On August 19, 1662, tortured with physical maladies and unpleasant mental conditions, Pascal's life came to an end when he ended up with convulsions and died in Paris at age 39.

The German Mathematician Christian Goldbach's Dilemma

Most children learning mathematics think that everything that is shown to be true is always true. Especially in mathematics, where so much of what is taught in schools is mechanical, very few people tend to question its veracity. Here's a little story you can tell your child about how a mathematician came up with an idea that seems to be correct but has never been proved. The story begins on June 7, 1742 with a letter by the German mathematician Christian Goldbach (1690–1764) to the famous Swiss mathematician Leonhard Euler (1707–1783), where he proposed an idea which until today has never been proved true and has never been proved untrue involving all even numbers. Goldbach's conjecture was that every even number greater than 2 can be expressed as the sum of two prime numbers.

You might want to have your child consider the following list of even numbers and their respective sum of prime numbers, and then have them continue it to convince themselves that it goes on — apparently — indefinitely.

Even numbers greater than 2	Sum of two prime numbers
4	$2 + 2$
6	$3 + 3$
8	$3 + 5$
10	$3 + 7$
12	$5 + 7$
14	$7 + 7$
16	$5 + 11$
18	$7 + 11$
20	$7 + 13$
$\cdots$	$\cdots$
48	$19 + 29$
$\cdots$	$\cdots$
100	$3 + 97$

You might mention as a point of history that there have been substantial attempts made at proving this conjecture by famous mathematicians: the German mathematician Georg Cantor (1845–1918) showed that the conjecture was true for all even numbers up to 1,000. In 1940 it was then shown to be true for all even numbers up to 100,000. By 1964, with the aid of a computer, it was extended to the number 33,000,000. In 1965 this was extended to 100,000,000; and then in 1980 to 200,000,000. Then in 1998 the German mathematician Jörg Richstein showed that Goldbach's conjecture was true for all even numbers up to 400 trillion.

In April 2012, Oliveira e Silva extended this conjecture to 4×10^{18}. Prize money of $1,000,000 has been offered for a proof of this conjecture. To date, this has not been claimed.[1] Along the way there were mistakes made when mathematicians tried to show this is true for all even numbers. But this conjecture that it holds true for all numbers still remains unsolved.

Unlike his other conjectures that seem to be correct, but have never been proved to be so, Goldbach made a further conjecture that was finally proved to be mistaken. In a letter to Euler on November 18, 1752, he stated the following:

Every odd number greater than 3 can be written as the sum of an odd number and twice a square number.

Here are a few examples of his conjecture.

Odd number	Sum of a prime number and the double of a square number (Goldbach considered 1 a prime number)
3	$= 1 + 2 \cdot 1^2$
5	$= 3 + 2 \cdot 1^2$
7	$= 5 + 2 \cdot 1^2$
9	$= 7 + 2 \cdot 1^2 = 1 + 2 \cdot 2^2$
11	$= 3 + 2 \cdot 2^2$
13	$= 5 + 2 \cdot 2^2 = 11 + 2 \cdot 1^2$
15	$= 7 + 2 \cdot 2^2 = 13 + 2 \cdot 1^2$
17	$= 17 + 2 \cdot 0^2$
19	$= 11 + 2 \cdot 2^2 = 17 + 2 \cdot 1^2$
21	$= 13 + 2 \cdot 2^2 = 19 + 2 \cdot 1^2$

---

[1] http://www.ieeta.pt/~tos/goldbach.html; Posamentier, A. S.; Lehmann, I.: *Mathematical Amazements and Surprises. Fascinating Figures and Noteworthy Numbers.* Afterword by Herbert Hauptman, Nobel Laureate. Amherst (New York), Prometheus Books, 2009, p. 226.

Euler replied on December 16, 1752, stating that he checked the first 1,000 odd numbers and found it to hold true. On April 3, 1753, Euler wrote Goldbach again, this time saying that he showed it held true for the first 2,500 odd numbers. However, in 1856, the German mathematician Moritz Stern (1807–1894) found that Goldbach's second conjecture did not hold true for the numbers 5,777 and 5,993, thereby rendering the conjecture a mistake. If one eliminates 0 as a possible square number, the Stern inspection of the first 9,000 odd numbers reveals the following numbers that also cannot fit the Goldbach conjecture: 17, 137, 227, 977, 1187, and 1493. To date, no other counterexamples have been found. Through Goldbach's recreational view of mathematics — despite mistakes — he stimulated much research in number theory. This sort of story will give your child a feeling that mathematics should be looked at in a critical manner and not always accept what appears to be a pattern — and isn't!

The Story of the Young German Mathematician
Carl Friedrich Gauss

Perhaps one of the cutest stories we can share with students is that of the young Carl Friedrich Gauss, who turned out to be one of the foremost German mathematicians of all time. Gauss was born Braunschweig, Germany on April 30, 1777. He was raised in a very poor family and got very little academic support from his parents. However, from a very early age he did show an unusually sharp photographic memory and an incredible talent with numbers. It is believed today that the inheritance of his genius came from his uncle Friedrich Benz, who also had shown some higher level of intelligence. Perhaps, the most famous story emerging from his childhood is one that happened at about age 10.

One day, when his teacher decided to keep the class busy, he assigned a problem to the class, which he believed would take them quite a bit of time to complete and thereby allowing him time to do his clerical work. The problem was to add up the natural numbers from 1 to 100. No sooner had he made the assignment that young Gauss placed his writing slate on the table, and claimed to have gotten the right answer. The teacher did not acknowledge this response and just asked the class to keep working. At the end of the allotted time, it turned out that young Gauss was the only student in the class to have gotten the right answer.

How is it that Gauss was able to get the answer so quickly? Had he known the answer in advance? Or did he have a clever method doing the addition? Rather than to add the numbers in the sequence that they are normally written, namely, $1 + 2 + 3 + 4 + \cdots + 97 + 98 + 99 + 100$, Gauss decided to add the numbers in pairs: $1 + 100 = 101, 2 + 99 = 101$, $3 + 98 = 101$, and realizing that there are 50 such pairs of 101, he simply multiplied $50 \times 101 = 5{,}050$, which gave him the required sum.

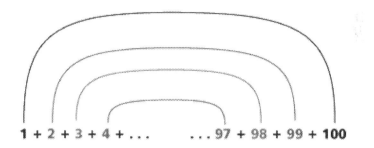

1 + 2 + 3 + 4 + ... 97 + 98 + 99 + 100

This clever procedure can also be used to generate a formula for finding the sum of any arithmetic sequence, that is, one where the difference between terms is the same. For an arithmetic sequence if n terms, the formula would then be: Sum $= n(\frac{a_1 + a_n}{2})$, where a_1 and a_n are the first and last terms, respectively. A nice little story like this can lead to a very useful mathematical technique!

Abraham Lincoln's Connection to the Famous Greek Mathematician Euclid

We would like to believe that mathematics ties in with all aspects of our lives and our studies. Here is one example of where mathematics ties in with what is taught in history. Perhaps Abraham Lincoln is one of the most famous presidents in our history. How can we tie in the life of Abraham Lincoln to mathematics? Abraham Lincoln claims to have not been at school more than six months in this entire life. He loved to read, and he read incredibly much. He did not go to law school and studied law on his own by reading the appropriate material and then passing the bar examination. During his law readings, he was fascinated by the word "demonstrate," which seems to have come up frequently in these readings. He claims to have gone to his father's house in order to understand that word better. There he found that to read the six books of Euclid's *Elements* — which he later claimed to be able to recite much of its contents from memory — provided that which he sought. (See Figure 6.5) Although the elements do not make for very exciting reading, they are the basis for the high school geometry course taught in the United States. Lincoln saw the work of Euclid in his book *Elements*, which proposes definitions that he believed were the foundations of knowledge, which then led to axioms that needed no proof. From this basis, Euclid proved numerous geometric propositions, which to a person of reason, such as Abraham Lincoln, became fact regardless of the person's belief system. It is well-known that during the early 1850s Abraham Lincoln carried a copy of Euclid's *Elements* in his saddle bags, and was known to dabble with geometric problems in his office.

Lincoln also invoked Euclid's thinking in the Lincoln–Douglas debates when he said: "If Judge Douglas will demonstrate somehow that this is popular sovereignty — the right of one man to make a slave of another, without any right in that other, or anyone else to object — demonstrate it as Euclid demonstrated propositions, there is no objection." This is just one example of how Abraham Lincoln, so frequently was infected by the logical thinking, demonstrated in Euclid's *Elements*.

Taking the effect of the elements step further towards the American high school curriculum, we have to acknowledge the Scottish mathematician Robert Simson (1687–1768), and thank him for taking a translated version of the *Elements* (Figure 6.6 shows the 1787 edition), and converting it into a course of study — albeit at the university level. The American mathematician Charles Davies (1798–1876) then adopted a consolidated work from Simson's *Geometry* by the French mathematician A. M. Legendre (1752–1833) to produce the first geometry text (Figure 6.7 shows an 1872 edition) that will have served as a model for American high schools for about 150 years.

We are often asked why we teach a year of geometry based on Euclid's elements? Aside from Lincoln, the English philosopher John Locke (1632–1704) believed strongly that the logical structure of the *Elements* enabled the thinking to distinguish between what was right and what was wrong. His thinking traveled across the Atlantic and infected Thomas Jefferson who in 1776 drafted the United States Declaration of Independence, which opens with "we hold these truths to be self-evident, that all men

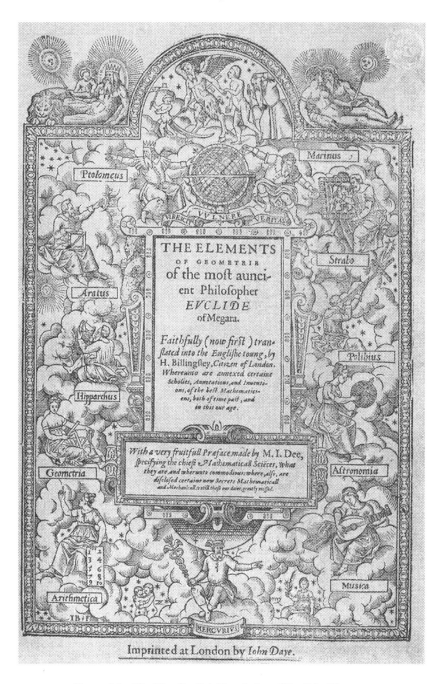

Figure 6.5: The First English Translation of Euclid's *Elements*

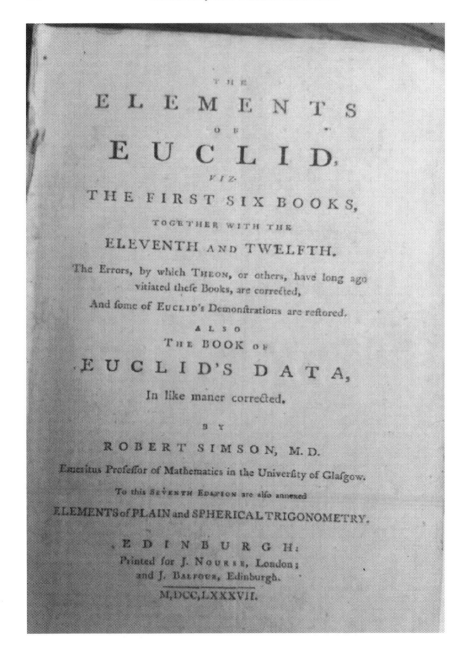

Figure 6.6:   Robert Simson's Translation of Euclid's *Elements*

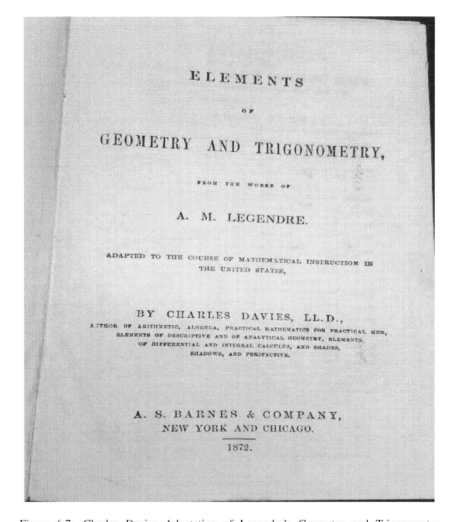

Figure 6.7: Charles Davies Adaptation of Legendre's Geometry and Trigonometry Textbook

are created equal." A strong advocate for studying mathematics, Jefferson adopted the thinking of Euclid and Locke with the notion of "self-evident." So, you can see how mathematics transcends itself and affects our thinking as our society developed. This is an important message to pass along to children of all ages, who are led to believe that mathematics is a separate subject and only applies to quantitative applications.

The Fabulous Fibonacci

Here is a story that will enchant most young people who might be curious about where our numbers actually came from and some of the fascinating things that could be done with them. We begin at the dawn of the 13th century, when mathematics and the European world began to get its modern image. This was largely due to the Italian mathematician Leonardo Pisano, best known as Fibonacci, who changed Western methods of calculation forever, and thereby, facilitating the exchange of currency and trade. He further presented mathematicians to this day with unsolved challenges, which are published through countless books and a quarterly journal published since 1963 by the Fibonacci Association.

It is good for young folks to see a connection between history and mathematics. The following will give you an opportunity to tie the two together going back to 12th-century. Leonardo Pisano, or Leonardo of Pisa, or as we said earlier, he is today known as Fibonacci, which is derived from the Latin "filius Bonacci," or a son of Bonacci, but it may more likely derive from "de filiis Bonacci," or family of Bonacci. He was born to Guilielmo Bonacci and his wife in the port city of Pisa, Italy around 1175, shortly after the start of construction of the famous bell tower, known today as the Leaning Tower of Pisa. These were turbulent times in Europe. The Crusades were in full swing, and the Holy Roman Empire was in conflict with the Papacy. The cities of Pisa, Genoa, Venice, and Amalfi, although frequently at war with each other, were maritime republics with specific trade routes to the Mediterranean countries and beyond. Pisa had played a powerful role in

Figure 6.8: Leonardo of Pisa (Fibonacci)

commerce since Roman times and even before as a port of call for Greek traders. Early on, it had established outposts for its commerce along its colonies and trading routes.

In 1192, Guilielmo Bonacci became a public clerk in the customs house for the Republic of Pisa, which was stationed in the Pisan colony of Bugia (today Bejaia, Algeria) on the Barbary Coast of Africa. Shortly after his arrival he brought his son, Leonardo, to join him so that the boy could learn the skill of calculating and become a merchant. This ability was significant since each republic had its own units of money and traders had to calculate monies due them. This meant determining currency equivalents on a daily basis. It was in Bugia that Fibonacci first became acquainted with the "nine Indian figures," 1, 2, 3, 4, 5, 6, 7, 8, 9, as he called the Hindu numerals, Hindu numerals, and "the sign 0 which the Arabs call zephyr". He declared his fascination for the methods of calculation using these numerals in the only source we have about his life story, the prologue to his most famous book, *Liber abaci* (with means book of calculations), which he wrote in 1202 and revised in 1228. This was the first time these Hindu Arabic numerals appeared in Europe. During his time away from Pisa, he received instruction from a Muslim teacher who introduced him to a book on algebra entitled *Hisâb al-jabr w'al-muqabâlah* by the Persian mathematician, al-Khowarizmi (ca. 780–ca. 850), which greatly influenced him. By the way, children might be fascinated that the name "algebra" comes from the title of this book.

During his lifetime, Fibonacci traveled extensively to Egypt, Syria, Greece, Sicily, and Provence, where he not only conducted business, but also met with mathematicians to learn their ways of doing mathematics. Indeed, Fibonacci sometimes referred to himself as "Bigollo" which could mean good-for-nothing or more positively, traveler. He may have liked the double meaning. When he returned to Pisa around the turn of the century, Fibonacci began to write about calculation methods with the Indian numerals for commercial applications in his book, *Liber abaci*. The volume consists largely of algebraic problems of "real world" situations that require more abstract mathematics. Fibonacci wanted to spread these newfound techniques to his countrymen.

It might be a good idea at this point to remind your child that during these times, there was no printing press, so books had to be handwritten by scribes, and if a copy was to be made, that too had to be handwritten. By the way, the printing press was first invented in Europe in the middle of the 15th century by Johannes Gutenberg, who lived in Mainz, Germany. Fibonacci's other books were *Practica geometriae* (1220), a book on the practice of geometry. It covers geometry and trigonometry with a rigor comparable to that of Euclid's work, and with ideas presented in proof form as well as numerical form, using these very convenient numerals. Here Fibonacci uses algebraic methods to solve geometric problems as well as the reverse. In 1225 he wrote *Flos* (on flowers or blossoms) and *Liber quadratorum* (or "Book of Squares"), a book that truly distinguished Fibonacci as a talented mathematician, ranking very high among number theorists. Fibonacci likely wrote additional works, however, there is no trace of them today. His book on commercial arithmetic, *Di minor guisa*, is lost as is his commentary on Book X of Euclid's *Elements*, which contained a numerical treatment of irrational numbers as compared to Euclid's geometrical treatment.

The confluence of politics and scholarship brought Fibonacci into contact with the Holy Roman Emperor Frederick the II (1194–1250) in the third decade of the century. Frederic, who had been crowned king of Sicily in 1198, king of Germany in 1212, and then crowned Holy Roman emperor by the Pope in St Peter's Cathedral in Rome (1220), had spent the years up to 1227 consolidating his power in Italy. He supported Pisa, then with a population of about 10,000, in its conflicts with Genoa at sea and with Lucca and Florence on land. As a strong patron of science and the arts, Frederick became aware of Fibonacci's work through the scholars at his court, who had corresponded with Fibonacci since his return to Pisa around 1200. These scholars included Michael Scotus who was the court astrologer, and to whom Fibonacci dedicated his book *Liber abaci*, Theodorus Physicus the court philosopher, and Dominicus Hispanus who suggested to Frederick that he meet Fibonacci when Frederick's court met in Pisa around 1225. The meeting took place as expected within the year.

Johannes of Palermo, another member of Frederick II's court, presented a number of problems as challenges to the great mathematician Fibonacci. Three of these problems were solved by Fibonacci. He provided solutions in *Flos*, which he sent to Frederick II.

One of the problems with which he was challenged we can explore here, since it doesn't require anything more than some elementary algebra. Remember that although these methods may seem elementary to us, they were hardly known at the time of Fibonacci, and so this was considered a real challenge. The problem was to find the perfect square that remains a perfect square when increased or decreased by 5.

Fibonacci found the number $\frac{41}{12}$ as his solution to the problem. To check this, we must add and subtract 5 and see if the result is still a perfect square.

$$\left(\frac{41}{12}\right)^2 + 5 = \frac{1681}{144} + \frac{720}{144} = \frac{2401}{144} = \left(\frac{49}{12}\right)^2$$

$$\left(\frac{41}{12}\right)^2 - 5 = \frac{1681}{144} - \frac{720}{144} = \frac{961}{144} = \left(\frac{31}{12}\right)^2$$

We have then shown that $\frac{41}{12}$ meets the criteria set out in the problem. Luckily the problem asked for 5 to be added and subtracted from the perfect square, for if he were asked to add or subtract 1, 2, 3, or 4 instead of 5, the problem could not have been solved.

Another problem, also presented in *Flos* is to solve the following:

> Three people are to share an amount of money in the following parts: $\frac{1}{2}$, $\frac{1}{3}$, and $\frac{1}{6}$. Each person takes some money from this amount of money until there is nothing left. The first person then returns $\frac{1}{2}$ of what he took. The second person then returns $\frac{1}{3}$ of what he took, and the third person returns $\frac{1}{6}$ of what he took. When the total of what was returned is divided equally among the three, each has his correct share, namely, $\frac{1}{2}$, $\frac{1}{3}$, and $\frac{1}{6}$. How much money was in the original amount and how much did each person get from the original amount of money?

Although none of Fibonacci's competitors could solve any of these problems, he gave as an answer 47 as the smallest amount, yet he claimed the problem was indeterminate.

The last mention of Fibonacci was in 1240, when he was honored with a lifetime salary by the Republic of Pisa for his service to the people, whom he advised on matters of accounting, often pro bono. We do not know exactly when Fibonacci died but it is believed that it was somewhere between 1240 and 1250 in Pisa.

Although Fibonacci was considered one of the greatest mathematicians at this time, his fame today is largely based on the book, *Liber abaci*. To appreciate his work let us consider this book as an example. This extensive

volume is full of very interesting problems based on the arithmetic and algebra, which Fibonacci had accumulated during his travels. The Hindu-Arabic
place-valued decimal system along with the use of the Arabic numerals was
widely copied and imitated, and, as noted, introduced into Europe. The book
was increasingly widely used for the better part of the next two centuries —
a best seller! Fibonacci begins *Liber abaci* with the following:

> "The nine Indian figures are:
>
> $$9\ 8\ 7\ 6\ 5\ 4\ 3\ 2\ 1.$$
>
> With these nine figures, and with the sign 0, which the Arabs
> call zephyr, any number whatsoever is written, as demonstrated
> below. A number is a sum of units, and through the addition
> of them the number increases by steps without end. First, one
> composes those numbers, which are from one to ten. Second,
> from the tens are made those numbers, which are from ten up to
> one hundred. Third, from the hundreds are made those numbers,
> which are from one hundred up to one thousand. . . . and thus,
> by an unending sequence of steps, any number whatsoever is
> constructed by joining the preceding numbers. The first place in
> the writing of the numbers is at the right. The second follows the
> first to the left."

Fibonacci used the term "Indian figures" for the Hindu numerals. Despite
their relative facility, these numerals were not widely accepted by merchants, who were suspicious of those who knew how to use them. They
were simply afraid of being cheated. We can safely say that it took the
same 300 years for these numerals to properly catch on, as it did for the
completion of the leaning tower of Pisa.

Interestingly, *Liber Abaci* also contains simultaneous linear equations.
Many of the problems that Fibonacci considers, however, were similar to
those appearing in Arab sources. This does not detract from the value of the
book, since it is the collection of solutions to these problems that makes the
major contribution to our development of mathematics. As a matter of fact,
a number of mathematical terms — common in today's usage — were first
introduced in *Liber abaci*. Fibonacci referred to "factus ex multiplicatione"
and from this first sighting of the word, we speak of the "factors of a multiplication." Incidentally, another example of words whose introduction into

the current mathematics vocabulary seems to stem from this famous book are the words "numerator" and "denominator."

The second section of *Liber abaci* includes a large collection of problems aimed at merchants. They relate to the price of goods, how to convert between the various currencies in use in Mediterranean countries, calculate profit on transactions, and problems that had probably originated in China.

Fibonacci was aware of a merchant's desire to circumvent the church's ban on charging interest on loans. So, he devised a way to hide the interest in a higher initial sum than the actual loan, and based his calculations on compound interest.

The third section of the book contains many problems such as:

> A hound whose speed increases arithmetically chases a hare whose speed also increases arithmetically, how far do they travel before the hound catches the hare.

> A spider climbs so many feet up a wall each day and slips back a fixed number each night, how many days does it take him to climb the wall.

> Calculate the amount of money two people have after a certain amount changes hands and the proportional increase and decrease are given.

There are also problems involving perfect numbers (which are those where the sum of their proper factors is equal to the number itself), problems involving the Chinese remainder theorem and problems involving the sums of arithmetic and geometric series. Fibonacci treats numbers such as $\sqrt{10}$ in the fourth section, both with rational approximations and with geometric constructions.

Some of the classical problems, which are considered recreational mathematics today, first appeared in the Western world in *Liber abaci*. This book is of particular interest to us, not only because it was the first publication in Western culture to use the Hindu numerals to replace the clumsy Roman numerals, or because Fibonacci was the first to use a horizontal fraction bar, but because it casually includes a recreational mathematics problem in chapter 12 that has made Fibonacci famous for posterity. This is the problem of the regeneration of rabbits.

The Famous Rabbit Problem: Figure 6.9 shows how this rabbit problem was originally stated (with the left-marginal note included):

Beginning	1
First	2
Second	3
Third	5
Fourth	8
Fifth	13
Sixth	21
Seventh	34
Eighth	55
Ninth	89
Tenth	144
Eleventh	233
Twelfth	377

"A certain man had one pair of rabbits together in a certain enclosed place, and one wishes to know how many are created from the pair in one year when it is the nature of them in a single month to bear another pair, and in the second month those born to bear also. Because the above written pair in the first month bore, you will double it; there will be two pairs in one month. One of these, namely the first, bears in the second month, and thus there are in the second month 3 pairs; of these in one month two are pregnant and in the third month 2 pairs of rabbits are born and thus there are 5 pairs in the month; in this month 3 pairs are pregnant and in the fourth month there are 8 pairs, of which 5 pairs bear another 5 pairs; these are added to the 8 pairs making 13 pairs in the fifth month; these 5 pairs that are born in this month do not mate in this month, but another 8 pairs are pregnant, and thus there are in the sixth month 21 pairs; to these are added the 13 pairs that are born in the seventh month; there will be 34 pairs in this month; to this are added the 21 pairs that are born in the eighth month; there will be 55 pairs in this month; to these are added the 34 pairs that are born in the ninth month; there will be 89 pairs in this month; to these are added again the 55 pairs that are both in the tenth month; there will be 144 pairs in this month; to these are added again the 89 pairs that are born in the eleventh month; there will be 233 pairs in this month. To these are still added the 144 pairs that are born in the last month; there will be 377 pairs and this many pairs are produced from the above-written pair in the mentioned place at the end of one year.

You can indeed see in the margin how we operated, namely that we added the first number to the second, namely the 1 to the 2, and the second to the third and the third to the fourth and the fourth to the fifth, and thus one after another until we added the tenth to the eleventh, namely the 144 to the 233, and we had the above-written sum of rabbits, namely 377 and thus you can in order find it for an unending number of months."

Figure 6.9:   A Translation of Fibonacci's Rabbit Problem from *Liber Abaci*

To see how this problem's situation would look on a monthly basis, consider the chart in Figure 6.10. If we assume that a pair of baby (B) rabbits matures in one month to become offspring-producing adults (A), then we can set up the following chart:

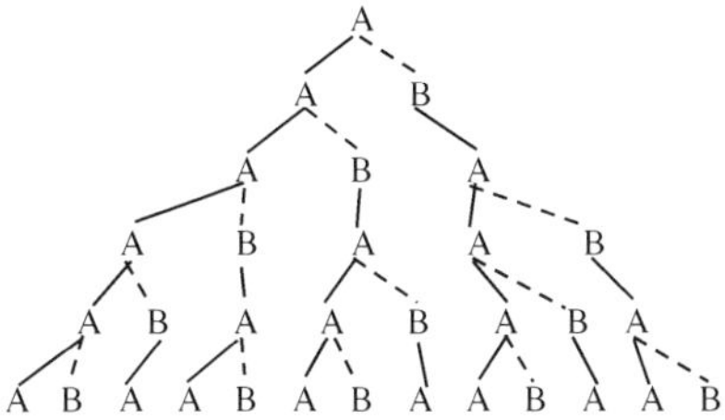

Month	Pairs	No. of Pairs of Adults (A)	No. of Pairs of Babies (B)	Total Pairs
Jan. 1		1	0	1
Feb. 1		1	1	2
Mar. 1		2	1	3
Apr. 1		3	2	5
May 1		5	3	8
June 1		8	5	13
July 1		13	8	21
Aug. 1		21	13	34
Sept. 1		34	21	55
Oct. 1		55	34	89
Nov. 1		89	55	144
Dec. 1		144	89	233
Jan. 1		233	144	377

Figure 6.10: A Chart Describing the Solution to Fibonacci's Rabbit Problem

This problem, which generated the sequence of numbers

1, 1, 2, 3, 5, 8, 13, 21, 34, 55, 89, 144, 233, 377, ...,

is known today as the *Fibonacci numbers*. At first glance there is nothing spectacular about these numbers beyond the relationship that would allow us to generate additional numbers of the sequence quite easily. We notice that every number in the sequence (after the first two) is the sum of the two preceding numbers. The Fibonacci sequence can be written in a way so that its recursive definition becomes clear: each number is the sum of the two preceding ones:

1

   1

1 + 1 = **2**

   1 + 2 = **3**

      2 + 3 = **5**

         3 + 5 = **8**

            5 + 8 = **13**

               8 + 13 = **21**

                  13 + 21 = **34**

                     21 + 34 = **55**

                        34 + 55 = **89**

                           55 + 89 = **144**

                              89 + 144 = **233**

                                144 + 233 = **377**

                                  233 + 377 = **610**

                                    377 + 610 = **987**

                                      610 + 987 = **1597** ...

The Fibonacci sequence is the oldest known (recursive) *recurrent* sequence. There is no evidence that Fibonacci knew of this relationship, but it is securely assumed that a man of his talents and insight knew the recursive relationship. It took another 400 years before this relationship appeared in print beyond his book.

Introducing the Fibonacci numbers

These numbers were not identified as anything special during the time Fibonacci wrote *Liber Abaci*. As a matter of fact, the famous German mathematician and astronomer, Johannes Kepler (1571–1630), mentioned these numbers in a 1611 publication[2] when he said that the ratios "as 5 is

---

[2]Maxey Brooke, "Fibonacci Numbers and Their History Through 1900," *Fibonacci Quarterly*, 2 (April 1964): 149.

Figure 6.11: François-Édouard-Anatole Lucas

to 8, so is 8 to 13, so is 13 to 21 almost." Centuries passed, and the numbers still went unnoticed. In the 1830s C. F. Schimper and A. Braun noticed the numbers appeared as the number of spirals of bracts on a pinecone. In the mid 1800's the Fibonacci numbers began to capture the fascination of mathematicians. They took on their current name ("Fibonacci numbers") from François-Édouard-Anatole Lucas (1842–1891), the French mathematician usually referred to as "Edouard Lucas," who later devised his own sequence by following the pattern set by Fibonacci. Lucas numbers form a sequence of numbers much like the Fibonacci numbers and also closely related to the Fibonacci numbers. Instead of starting with 1, 1, 2, 3, 5, . . ., Lucas began his sequence with 1, 3, 4, 7, 11, ...

At about this time the French mathematician, Jacques-Philippe-Marie Binet (1786–1856), developed a formula for finding any Fibonacci number given its position in the sequence. That is, with Binet's formula we can find the 118th Fibonacci number without having to list the previous 117 numbers. The formula is: $F_n = \frac{1}{\sqrt{5}}\left[\left(\frac{1+\sqrt{5}}{2}\right)^n - \left(\frac{1-\sqrt{5}}{2}\right)^n\right]$, where F_n is the nth Fibonacci number. Although this formula looks quite complicated, it isn't as bad as it looks. However, it is very powerful because it allows us to find any Fibonacci number without going through the whole list to arrive at it. The Fibonacci numbers are probably the most famous sequence of numbers in all of mathematics, since they appear in just about every aspect of our experiences.

Still your child may ask, what is so special about these numbers? Let us just begin to scratch the surface by simply inspecting this famous Fibonacci number sequence and some of the remarkable properties it has.

As we did above, we will use the symbol F_7 to represent the 7th Fibonacci number, and F_n to represent the nth Fibonacci number. Consider the first 30 Fibonacci numbers shown in Figure 6.12.

With all these lovely relationships embracing the Fibonacci numbers, there must be a simple way to get the sum of a specified number of these Fibonacci numbers. A simple formula would be helpful as opposed to actually adding all the Fibonacci numbers to a certain point. To derive such a

$$F_1 = 1 \qquad\qquad F_{16} = 987$$

$$F_2 = 1 \qquad\qquad F_{17} = 1597$$

$$F_3 = 2 \qquad\qquad F_{18} = 2584$$

$$F_4 = 3 \qquad\qquad F_{19} = 4181$$

$$F_5 = 5 \qquad\qquad F_{20} = 6765$$

$$F_6 = 8 \qquad\qquad F_{21} = 10946$$

$$F_7 = 13 \qquad\qquad F_{22} = 17711$$

$$F_8 = 21 \qquad\qquad F_{23} = 28657$$

$$F_9 = 34 \qquad\qquad F_{24} = 46368$$

$$F_{10} = 55 \qquad\qquad F_{25} = 75025$$

$$F_{11} = 89 \qquad\qquad F_{26} = 121393$$

$$F_{12} = 144 \qquad\qquad F_{27} = 196418$$

$$F_{13} = 233 \qquad\qquad F_{28} = 317811$$

$$F_{14} = 377 \qquad\qquad F_{29} = 514229$$

$$F_{15} = 610 \qquad\qquad F_{30} = 832040$$

Figure 6.12:   The First 30 Fibonacci Numbers

formula for the sum of the first n Fibonacci numbers, we will use a technique that will help us generate a formula. From the definition of the Fibonacci numbers, we can write that symbolically as $F_{n+2} = F_{n+1} + F_n$, where $n > 1$. This can be rewritten as $F_n = F_{n+2} - F_{n+1}$. By substituting increasing values for n we get the following:

$$F_1 = F_3 - F_2$$

$$F_2 = F_4 - F_3$$

$$F_3 = F_5 - F_4$$

$$F_4 = F_6 - F_5$$

$$\vdots$$

$$F_{n-1} = F_{n+1} - F_n$$

$$F_n = F_{n+2} - F_{n+1}$$

By adding these equations, you will notice that there will be many terms on the right side of the equations that will disappear (because their sum is zero — since you will be adding and subtracting the same number). What will remain on the right side will be $F_{n+2} - F_2 = F_{n+2} - 1$. On the left side we have the sum of the first n Fibonacci numbers: $F_1 + F_2 + F_3 + F_4 + \cdots + F_n$, which is what we are looking for.

Therefore, we get the following: $F_1 + F_2 + F_3 + F_4 + \cdots + F_n = F_{n+2} - 1$, which says that the sum of the first n Fibonacci numbers is equal to the Fibonacci number two further along the sequence minus 1. This can also be written symbolically as $\sum_{i=1}^{n} F_i = F_{n+2} - 1$.

Just for entertainment, and to entice you a bit to perhaps enjoy the Fibonacci numbers in greater detail consider the following illustrations:

The sum of *any ten* consecutive Fibonacci numbers is divisible by 11. We could convince ourselves that this may be true by considering some randomly chosen examples. Take, for example, the sum of the following ten consecutive Fibonacci numbers: $13 + 21 + 34 + 55 + 89 + 144 + 233 + 377 + 610 + 987 = 2{,}563$, which is divisible by 11, since $11 \times 233 = 2{,}563$. We could repeat this for any other sum of 10 consecutive Fibonacci numbers such as the sum of the ten consecutive Fibonacci numbers from F_{21} to F_{30}, which is $2{,}160{,}598 = 11 \times 196{,}418$.

One way to go about convincing yourself of the truth in this "conjecture" is to keep on taking the sum of groups of ten consecutive Fibonacci numbers and checking to see if the sum is a multiple of 11. You could also try to prove the statement, mathematically. Listing the remainders of the first few Fibonacci numbers upon dividing by 11, we have

$$1, 1, 2, 3, 5, 8, 2, 10, 1, 0, \underline{1, 1, 2, 3, 5, 8, 2, 10, 1, 0}, \ldots.$$

We see that the remainders repeat in cycles of length 10. Since it is the remainder upon dividing a number by 11 that determines its divisibility by 11, all we have to do is check that in adding any 10 consecutive numbers in the sequence 1, 1, 2, 3, 5, 8, 2, 10, 1, 0, 1, 1, 2, 3, 5, 8, 2, 10, 1, 0, ... we get a sum divisible by 11. We can check this as follows. Since the cycle of this sequence is of length exactly 10, adding any 10 consecutive numbers in this sequence will always come out to adding the 10 numbers — 1, 1, 2, 3, 5, 8, 2, 10, 1, 0 — in a cycle.

Imagine these 10 numbers arranged clockwise in that order around a circle, with the sequence above obtained by traveling around the circle over and over. Then you can see that any numbers missed at the beginning of a cycle — if the sum is started somewhere in the interior of the cycle — are regained from the next cycle; for example, the sum $5 + 8 + 2 + 10 + 1 + 0 + \underline{1 + 1 + 2 + 3}$. This is because no matter where you start counting on the circle, counting 10 numbers clockwise around the circle will amount to counting all 10 numbers, because that's exactly how many numbers there are.

These 10 numbers have sum 33, which is indeed divisible by 11.

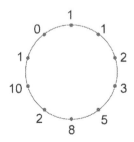

Figure 6.13

Aside from the fact that Fibonacci is largely responsible for the use of our decimal number system, having introduced it to the Western world in 1202, he is primarily remembered today for the ubiquitous numbers that bear his name. If this brief presentation of the Fibonacci numbers interested you, and you would like to see more applications that you can share with your child, we recommend the following book: *The Fabulous Fibonacci Numbers*, A. S. Posamentier and I. Lehmann, (Prometheus Books, 2007).

The Most Prolific Mathematician of All Time, the Swiss Mathematician Leonhard Euler

Children often wonder from where our many mathematical symbols stem. The answer is quite simple. Perhaps one of the most prolific mathematicians of all time, the Swiss mathematician Leonhard Euler, through his many writings introduced a significant number of the symbols that we frequently use in mathematics today. Although the Greek letter π was first used by the Welsh mathematician William Jones (1675–1749), it was Euler who through his many publications popularized the symbol to represent the ratio of a circle's circumference to its diameter. Although this may be a little bit beyond the scope of young children, they may still be fascinated to know that it was Leonhard Euler who first used the Greek letter Σ to represent a summation, and the letter i to represent the imaginary number $\sqrt{-1}$. He was also the first to use the letter e to represent the natural logarithm, which is approximately equal to $2.71828\ldots$, and that allowed him to set up the famous Euler identity $e^{i\pi} + 1 = 0$, which uses all of these symbols. Euler also introduced the concept of the function and was the first to write it as $f(x)$. We can thank Euler for the modern notation of the trigonometric functions. Even the way we today label geometric figures, such as a triangle, whose vertices are marked the letters A, B, and C, and the notation for the sides opposite these vertices using the lowercase letters a, b, and c, stems from Euler's writings. Although we introduced Euler through his providing us with our mathematical language, this resulted from the many volumes that he has written during his 76-year life span. Yet we must also acknowledge the multitude of innovations that Euler introduced in mathematics. But first, let us get a brief view into his life's history because it is not only unique but also from a historical point quite extraordinary.

Leonhard Euler (Portrait by Jakob Emanuel Handmann — 1753)

Leonhard Euler was born on April 15, 1707 in Basel, Switzerland. His father was a pastor and his mother was the daughter of a pastor and he was one of four children. When Leonhard was a small child his family moved to the town of Riehen, where he lived until age 13 at which time he moved back to Basel to live with his maternal grandmother. There he enrolled as a student at the University of Basel, where in 1723 he received a Master's degree in philosophy. It was during that time that his father's friend Johann Bernoulli gave him private lessons in mathematics and discovered his unique talents in this field. As a result, Bernoulli convinced Leonhard's father that his son should pursue a study of mathematics rather than his father's wish that he also become a pastor. Euler completed his dissertation on the propagation of sound in 1726, but was unsuccessful in obtaining a faculty position at the University of Basel. In 1727, Euler accepted a position in the mathematics department at the Academy of Russia. This Academy was eager to recruit scholars from other European countries and was chiefly interested in research rather than teaching students. During this time, Euler mastered the Russian language and also took on an additional job as a medic in the Russian Navy. In 1731 he was promoted to professor of physics and two years later Euler headed the mathematics department there as well, a position vacated by Daniel Bernoulli (1700–1782), Johann Bernoulli's son.

In 1734 Euler married Katharina Gsell, with whom he had 13 children of which only 5 survived childhood. In 1738, apparently through a protracted fever and overwork, Euler lost sight in his right eye. This did not distract him from continuing his work as energetically as previously. In 1741, the instability in Russia motivated Euler to take on a faculty position at the Berlin Academy, where he stayed for the next 25 years and wrote over 380 articles. Also, during his Berlin years, he wrote two very famous books. One book on the topic of functions, *Introducto in analysin infinitorum*, was published in 1748, and in 1755 a book on differential calculus entitled *Institutiones calculi differentialis*. In 1770 he completed a book entitled *Institutiones calculi integralis*. These last two books provide many formulas for differentiation and integration, which comprise much of our modern-day calculus course.

In 1766 Euler accepted an invitation from Catherine II to return to Russia, where soon after his arrival in St. Petersburg a cataract formed in his remaining good eye, which left him totally blind for the rest of his life. Amazingly, this did not reduce his productivity, largely because of his uncommon memory and unusual ability to do mental calculations. He actually claimed that losing sight enabled him to have fewer distractions in his work. With his scribes he was able to produce even more work than previously. Although he was not seen as a teacher of mathematics, he did have a great influence on mathematics education in Russia. Euler touched so many areas of mathematics that it would take volumes to summarize all of his work. However, we will cite two results of his genius that still remain somewhat popular today even beyond mathematical circles. Actually, youngsters might enjoy applying the findings of Euler's work as exhibited in these two examples.

First, there is a famous problem in mathematics that stems from an age-old conundrum that fascinated folks in Europe for many years. Although we discussed this in chapter 5, we offer again a little bit of a historical background so that your child can become fascinated by the problem that faced generations of Europeans.

In the eighteenth century and earlier, when walking was the dominant form of local transportation, people would often count particular kinds of objects they passed. One such was bridges. Through the eighteenth century the small Prussian city of Königsberg (today called Kaliningrad, Russia), located where the Pregel River forms two branches, was faced with a recreational dilemma: Could a person walk over each of the seven bridges *exactly once* in a continuous walk through the city? The residents of the city had this as a recreational challenge, particularly on Sunday afternoons. Since there were no successful attempts, the challenge continued for many years. As we indicated earlier, this problem provides a wonderful window into the field today known as *networks*, which is also referred to as *graph theory*, an extended field of geometry, which in the simple level could be within the scope of a child. This problem gives us a nice introduction into the subject. You might want to revert back to chapter 5 to see the mathematics behind this amazing problem.

Another phenomenon that can be easily presented to children, albeit slowly and carefully is, one that Euler established for any convex polyhedron, namely, that the relationship between the vertices (V), edges (E), and faces (F) satisfies the following equation: $V + F = E + 2$, which is known as the *Euler formula*. It might be fun to have your child verify this formula with any convex polyhedron you may have available. Your child might begin with the five regular polyhedra shown in Figure 6.14.

On September 18, 1783, during a discussion of planetary motion, Euler collapsed from a brain hemorrhage and died a few hours later. This most prolific mathematician and scientist was also heavily involved in areas beyond pure mathematics, such as cartography, physics, and astronomy just to name a few. We still remember Euler today as having contributed more volumes of work to mathematics than any other mathematician in history. A large

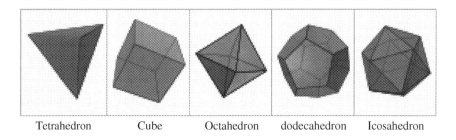

| Tetrahedron | Cube | Octahedron | dodecahedron | Icosahedron |

Figure 6.14: The Five Regular Polyhedra

number of mathematical objects and topics are named in honor of Leonhard Euler. In fact he made so many pioneering contributions to several branches of mathematics that some theorems were deliberately attributed to the first mathematician to have discovered them after Euler, in order to avoid naming everything after Euler.

The British Mathematician Charles Babbage Who Produced the World's First Calculator

Every child today is aware of the electronic calculators that are everywhere to be found, especially as an application. In other words, in today's modern world the calculator and the computer are pretty much taken for granted. However, your child might be interested to determine from where the concept of a calculator, or computer as it was originally known, emanated. This honor of having first developed a machine that does calculation belongs to the English mathematician Charles Babbage (Figure 6.15), who was born in London on December 26, 1791. Again, little history can only serve to enrich a discussion about mathematics.

He was the son of Benjamin Babbage who was a London banker. Charles Babbage was educated mostly at home, since he was frequently ill. Even in

Figure 6.15: Charles Babbage

these early days he developed an early love for mathematics. In 1810, he was accepted to study mathematics at Trinity College of Cambridge University, where in a short time he found himself more advanced than his instructors in mathematics. This prompted him to join a group of students also disappointed with the level of instruction. This group, called the Analytical Society, was devoted to exploring more advanced issues in mathematics. During this time, he and his colleagues were disturbed by the inaccuracy of the number of logarithm tables. He felt better about calculating these values himself, which was the initial motivation to develop a machine that could do that task accurately. In 1817 he received his M.A degree from Cambridge University. Soon thereafter, he worked there as a lecturer of mathematics at Cambridge University.

By 1816 he was elected a fellow of the Royal Society, and thereupon participated in the founding of the Royal Astronomical Society in 1920. This was about the time where his interest began to take him in the direction of developing a calculating machine, whose initial interest had been lurking in his mind for some time. In 1821 Babbage developed a Difference Engine No. 1 (Figure 6.16). The work on the conceived machine was compensated by governmental funds. Babbage did not get along well with the funders, since many of the questions seem to have been, by his measure, ridiculous. Eventually, funding was withdrawn, because it took so long to reach a workable model. Despite the lack of financial support, his model was completed in 1832 and was able to assist in compiling mathematical tables. Yet, he was displeased with the limitations of this machine, and so he began to develop one that could do a variety of calculations.

By 1856, he developed a more sophisticated calculating machine, the Analytical Engine, which could be considered a forerunner of today's computers. This machine was to be capable of doing arithmetic operations based on a memory recorded on punch cards, which is clearly how the early computers were functioning. After that time, he did not complete any of his further designs. In 1843, George Scheutz (1785–1873), a Swedish inventor was able to construct the Difference Machine based on Babbage's design. A second version of this machine, about the size of a piano, was developed in 1853. This machine was able to print logarithmic tables, trigonometric functions, actuarial tables, and astronomical tables with incredible accuracy.

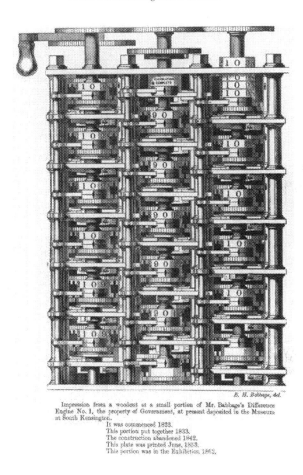

R. H. Babbage, del.

Impression from a woodcut of a small portion of Mr. Babbage's Difference
Engine No. 1, the property of Government, at present deposited in the Museum
at South Kensington.
It was commenced 1823.
This portion put together 1833.
The construction abandoned 1842.
This plate was printed June, 1853.
This portion was in the Exhibition 1862.

Figure 6.16:   Difference Engine No. 1 (Woodcut, 1853)

Babbage continued at Cambridge University, where he held the Lucasian
Chair of Mathematics from 1828 to 1839, but never presented a lecture. Babbage was a very active member of the intellectual society and supported
research in many scientific areas, which were then used by the British and
American governments. In 1837, Babbage designed a successor to the second version of the Difference Engine and called it the Analytical Engine.
This was a general-purpose computer, whose design regarding memory was
the forerunner of the electronic computers that followed years later. The
conceived memory was to hold 1000 numbers comprised of 40 decimal

Figure 6.17:   Analytical Engine

digits each. The machine was intended to perform the four arithmetic operations as well as square root extraction. The programming language used
was analogous to the modern-day assembly languages. Punch cards were
used, one for arithmetic operations, one for numerical constants and one for
transferring numbers from storage to the arithmetic unit. Unfortunately, the
machine was never successfully completed to the level Babbage desired,
and ran only a few tasks with some obvious errors. It is sad to note that
Babbage died a frustrated man, since his visions were never fully realized,
which he blames on the government's failure to provide the proper financial
support. After Babbage's death on October 18, 1871, his work was continued by his son Henry Prevost Babbage. In fact, he had to do much of his
own funding to continue his work. In 1910, Henry Babbage constructed his
version of the Analytical Engine, which was not programable and had no
storage (Figure 6.18).

Your child might be also interested to know that Babbage's legacy
includes many diverse contributions such as having compiled reliable
actuarial tables, having assisted in establishing the modern British postal

Figure 6.18: Henry Babbage's Analytical Engine Mill, built in 1910, on exhibit in the London Science Museum

system. He also invented a speedometer, occulting lights for lighthouses and a locomotive cow catcher. So here we have the initiator of our computer world struggling to make a machine that was motivated by what he saw as something desperately needed to correct earlier accepted erroneous information.

There are many sources from which to select other such historical stories that can be used to enhance mathematics. Children should know that mathematics should not be isolated from the rest of their studies, but can be easily combined with other subjects, such as in this case history.

One such possible source to consider is ***Math Makers: The Lives and Works of 50 Famous Mathematicians***, by A. S. Posamentier and C. Spreitzer (Prometheus, 2019).

Summing-up: A Note to Parents

We hope that the preceding chapters have provided some support in helping you take a more active role in your child's mathematics education. Perhaps your understanding of mathematics has been enhanced, or the importance of learning mathematics highlighted. Hopefully some of the offered strategies for overcoming challenges as you embark on taking a more proactive role in supporting mathematics education in the home will be helpful.

Chapters 1 and 2 set the stage for helping children learn mathematics by identifying challenges that you can expect to encounter, and by providing strategies for addressing these challenges. While these are psychological factors, and not mathematics concepts, anticipating them and addressing them can create the opportunity to improve children's learning. Behavioral strategies can be used to create a predictable environment in which children may feel they have some control over their learning. Addressing mathematics anxiety, which, unfortunately, is quite prevalent, can free children from their fears, and consequently permit them to learn new material.

We are all aware of the changes in mathematics instruction in the elementary grades. Needless to say, these new approaches are also introduced and used in more advanced grades, which is why it is essential for children to learn them early on. Chapter 3 presented a detailed review of how mathematics should be seen at the basic levels of arithmetic and beyond, consistent with the principles outlined in current national standards. In particular, the need for understanding mathematics concepts — the 'why' and not only the 'how' — has been developed here.

All too often, children ask why they have to learn mathematics? When will it be useful? Chapter 4 provided familiar, interesting, and intriguing answers to those questions. You will find unusual, and often entertaining, applications of mathematics that will allow you and your children to appreciate that learning mathematics can be useful. Clearly, we have merely scratched the surface when it comes to exhibiting everyday mathematical applications. We would, however, recommend a publication that presents a much broader picture with many, many more examples: *The Mathematics of Everyday Life* (A. S. Posamentier, and C. Spreitzer, Prometheus Books, 2018). Naturally, parents need to be very careful to select applications that are meaningful to their children as well as appropriate for their grade level.

Enticing mathematics enrichments can be found at all grade levels. Chapter 5 offered a variety of rather simple entertainments that parents can use at home to show their children some amazing aspects to mathematics. It is not uncommon that children will then promptly go and show off their new-found ideas to their friends, thereby popularizing mathematics beyond the classroom. This should have favorable results when they return to the curricula requirements presented in the normal school day.

The history of mathematics and its development over the centuries lends itself to stories that not only demonstrate some mathematical peculiarities, but also reveal interesting facts about prominent mathematicians. Chapter 6 provided stories that can be shared with your children during leisure times, to demonstrate that mathematics is much more than merely memorizing procedures and number facts, but also has an interesting dynamic life of its own.[1]

This book recognizes the unique role of each parent in their children's learning of mathematics. It provides behavioral and content ideas for assuming your role as a key figure in the education of your children, as well as resources for creating an effective mathematics-based relationship with them. We look forward to your success in this important endeavor.

---

[1] For more enticing stories about famous mathematicians see: Posamentier, Alfred S. and Christian Spreitzer, *Math Maker: The lives and Works of 50 Famous Mathematicians* (Prometheus Books, 2019).

Printed in the United States
By Bookmasters